Marc Reinhardt

Linear Estimation in Interconnected Sensor Systems with Information Constraints

Karlsruhe Series on Intelligent Sensor-Actuator-Systems

Volume 16

ISAS | Karlsruhe Institute of Technology
Intelligent Sensor-Actuator-Systems Laboratory

Edited by Prof. Dr.-Ing. Uwe D. Hanebeck

Linear Estimation in Interconnected Sensor Systems with Information Constraints

by
Marc Reinhardt

KIT Scientific Publishing

Dissertation, Karlsruher Institut für Technologie (KIT)
Fakultät für Informatik, 2014

Impressum

KIT Scientific Publishing

Karlsruher Institut für Technologie (KIT)
KIT Scientific Publishing
Straße am Forum 2
D-76131 Karlsruhe

KIT Scientific Publishing is a registered trademark of Karlsruhe Institute of Technology. Reprint using the book cover is not allowed.

www.ksp.kit.edu

Print on Demand 2015

ISSN ISSN 1867-3813
ISBN 978-3-7315-0342-2
DOI 10.5445/KSP/1000045577

Linear Estimation in Interconnected Sensor Systems with Information Constraints

zur Erlangung des akademischen Grades eines

Doktors der Ingenieurwissenschaften

von der Fakultät für Informatik
des Karlsruher Instituts für Technologie (KIT)

genehmigte

Dissertation

von

Marc Reinhardt

aus Kassel

Tag der mündlichen Prüfung: 19.12.2014

Erster Gutachter: Prof. Dr.-Ing. Uwe D. Hanebeck

Zweiter Gutachter: Prof. Sanjeev R. Kulkarni (Ph.D.)

Acknowledgment

Crafting a thesis is seldom the result of an individual effort. Over the last three years I had the chance to meet and collaborate with many very intelligent, knowledgeable, and kind researchers that have fundamentally impacted the work you are currently reading. Only thanks to many fruitful discussions in meetings and on conferences was I able to develop an advanced understanding of linear estimation theory, which finally allowed me to compile this thesis.

First of all I want to thank my adviser Uwe D. Hanebeck for giving me the opportunity to work in this captivating area. His valuable inputs always reminded me to consider the "bigger picture", which ultimately allowed me a better and more fundamental understanding of my research topics. I am also especially grateful to Sanjeev R. Kulkarni. His kind attitude and intelligent questions made my stay in Princeton not only a great experience from a research perspective but also a wonderful time in itself. In particular, I am honored that he spent his very limited time and served as my co-adviser.

Indeed, day-to-day research work meant for me to overcome mathematical problems. I am really fortunate that I could always count on my helpful and competent colleagues from the ISAS laboratory when I was stuck. Special thanks are directed to Benjamin Noack, who served as a great mentor and with whom I collaborated in many very interesting scientific works. I would also like to thank Marcus Baum and Jörg Fischer explicitly with whom I had quite some fruitful scientific and also political discussions over the last years.

Moreover, I was very lucky to be given the opportunity to attend several conferences and to meet internationally renowned scientists that had already achieved great results in linear estimation theory and whose ideas greatly inspired my works. In particular, I would like to thank Felix Govaers, Joris Sijs, and Jiří Ajgl for their valuable inputs, interesting exchanges of ideas, and our successful common works. A very special thanks also goes to Shang Shang who supported me before and during my research stay and with whom I have become friends.

In fact, many of my works originated from the cooperation with scientists from all over the world. Such collaborations were only possible as I received generous support from the KIT, the Karlsruhe House of Young Scientists, and the Research Training Group 1194. Apart from the funding, I also greatly benefited from the supportive environment that these institutions provided. In particular, the discussions with colleagues from the Research Training Group were really inspiring and gave me a much better understanding of the challenges in sensor actuator networks.

Last but not least, I would like to thank all the people outside academia who have supported me during the last three years. I am in particular grateful to my family and to Saskia for always standing behind me and for releasing me as good as possible from all routine work during stressful times.

Thank you!

Karlsruhe, January 2015 Marc Reinhardt

Contents

Notation

General Conventions

x	Scalar (lowercase)
$\mathbf{x}$	Random variable (bold, lowercase)
$\underline{x}$	Vector (underlined, lowercase)
$\underline{\mathbf{x}}$	Random vector (bold, underlined, lowercase)
$\mathbf{A}$	Matrix (bold, uppercase)
$(.)^s$	Quantity of sensor s
$(.)_k$	Quantity at time step k
$\mathcal{A}$	Set (calligraphic, uppercase)
$\mathbf{A} > \mathbf{0}$	Matrix $\mathbf{A}$ is positive definite
$\mathbf{A} \geq \mathbf{0}$	Matrix $\mathbf{A}$ is positive semidefinite
$\mathbf{A} > \mathbf{B}$	Matrix $(\mathbf{A} - \mathbf{B})$ is positive definite

Operators

$(\mathbf{A})^\top$	Matrix transpose of $\mathbf{A}$
$(\mathbf{A})^+$	Moore-Penrose pseudoinverse of $\mathbf{A}$
$(\mathbf{A})^{-1}$	Inverse of $\mathbf{A}$
$\mathrm{tr}\{\mathbf{A}\}$	Trace of $\mathbf{A}$
$\mathrm{rank}\{\mathbf{A}\}$	Rank of $\mathbf{A}$
$\mathrm{vec}\{\mathbf{A}\}$	Vector operator of $\mathbf{A}$
$\mathrm{eig}\{\mathbf{A}\}$	Set of eigenvalues of $\mathbf{A}$
$\mathrm{diag}(\mathbf{A})$	Vector with diagonal elements of $\mathbf{A}$
$\mathrm{blkdiag}(\mathbf{A})$	Matrix-vector of block diagonals of $\mathbf{A}$
δ	Kronecker delta function
$\mathcal{J}\{\cdot\}$	Optimization or cost function
$p(\underline{\mathbf{x}})$	Probability density func. of random variable $\underline{\mathbf{x}}$
$\Pr\{\underline{\mathbf{x}} = a\}$	Probability of event a
$\mathrm{E}\{\underline{\mathbf{x}}\}$	Expected value of $\underline{\mathbf{x}}$
$\mathrm{E}\{\underline{\mathbf{x}}\vert\underline{\mathbf{y}}\}$	Expected value of $\underline{\mathbf{x}}$ conditioned on $\underline{\mathbf{y}}$
$\mathcal{A} \cup \mathcal{B}$	Union of $\mathcal{A}$ in $\mathcal{B}$
$\mathcal{A} \cap \mathcal{B}$	Intersection of $\mathcal{A}$ in $\mathcal{B}$
$\mathcal{A} \backslash \mathcal{B}$	Relative complement of $\mathcal{B}$ in $\mathcal{A}$
$\mathcal{E}(\mathbf{C})$	Covariance ellipsoid of $\mathbf{C}$ with mean $\underline{0}$

Symbols

$\mathbf{I}$	Identity matrix
$\mathbf{1}$	Matrix with identities $(\mathbf{I} \;\; \ldots \;\; \mathbf{I})$
$\underline{0}$	Vector where all entries are zero
$\mathbf{0}$	Matrix where all entries are zero
$\mathbb{N}_0$	Set of natural number including zero
$\mathbb{N}$	Set of natural number excluding zero
$\mathbb{R}$	Set of real number
$\mathbb{R}^n$	n-dimensional vector space over the real numbers
$\emptyset$	Empty set
$\square$	End of proof

Conventions for Variables

$\mathcal{S}$	Set of sensor indices
$\mathcal{K}$	Set of considered time indices
$\mathcal{I}$	Set of noise term indices
$\underline{\mathbf{x}}_k$	True state at time k
$\underline{\mathbf{z}}_k$	Measurement obtained by sensor at time k
$\underline{\mathbf{w}}_k$	Process noise at time k
$\underline{\mathbf{v}}_k$	Measurement noise at time k
$\mathbf{A}_k$	State transition matrix at time k
$\mathbf{H}_k^s$	Measurement matrix of sensor s at time k
$\mathbf{Q}_k$	Process noise covariance at time k
$\mathbf{R}_k^s$	Measurement noise cov. of sensor s at time k
$\mathbf{C}^z$	Measurement capacity of sensor network
$\hat{\mathbf{C}}^z$	Hypothesis about measurement capacity
$\mathbf{K}$	filter gain for measurement
$\mathbf{L}$	filter gain for predicted estimate
$\mathbf{F}$	Fusion gain
$\mathcal{N}(\underline{x}, \mathbf{C})$	Gaussian distribution with mean $\underline{x}$ and cov. $\mathbf{C}$

Abbreviations

MSE	Mean Squared Error
MMSE	Minimum Mean Squared Error
LMMSE	Linear Minimum Mean Squared Error
KF	Kalman Filter
DKF	Distributed Kalman Filter
HKF	Hypothesizing Distributed Kalman Filter
LRKF	Linear Regression Kalman Filter
EKF	Extended Kalman Filter
UKF	Unscented Kalman Filter
FKF	Federated Kalman Filter
CI	Covariance Intersection

Zusammenfassung

Die Verarbeitung von Informationen aus unterschiedlichen Quellen zum Ableiten von Schlussfolgerungen fasziniert Wissenschaftler und Ingenieure gleichermaßen und viel spricht dafür, dass die Bedeutung der zugehörigen Theorie in einer immer stärker vernetzten Welt zukünftig noch zunehmen wird. Die mathematischen Grundlagen zur statistischen Evaluation und Interpretation solch unsicherer Informationen werden in der stochastischen Informationsverarbeitung entwickelt. Die Schätztheorie in Sensornetzen konstituiert einen wichtigen Teilbereich, der sich mit der verteilten Verarbeitung von Informationen mit beschränktem lokalem Wissen beschäftigt.

Während gemäß ursprünglicher Vorstellungen rechenschwache Miniatursysteme unter Berücksichtigung limitierter Batteriekapazitäten ein Netzwerk zur Überwachung räumlich verteilter Phänomene bilden sollten, hat sich in den letzten Jahren gezeigt, dass die Ideen und Methoden in diversen modernen aber auch klassischen Technologien eingesetzt werden können. Die Integration von Sensordaten in Smartphones und die Distribution komplexer Wettersimulationen sind dabei nur zwei Beispiele für mögliche Anwendungsgebiete. Bemerkenswerterweise erfordert die steigende Komplexität selbst in traditionell zentralen Systemen eine Dekomposition von Operationen in funktionell unabhängige Teilsysteme, die ähnlichen Ansprüchen wie Knoten in klassischen Sensornetzen gerecht werden müssen.

Das Ziel dieser Dissertation ist die Entwicklung verteilter Algorithmen zur effizienten Schätzung eines Zustands, der ausschließlich über unsichere Informationen zugänglich ist. Der bearbeitete Themenkomplex wird in der Literatur auch unter den Begriffen Target Tracking, verteilte und dezentrale Schätzung sowie Multisensor Datafusion behandelt. Die zusätzlichen Probleme gegenüber der zentralen Verarbeitung manifestieren sich in der unsicheren (häufig drahtlosen) Kommunikation, limitierten Wissen in lokalen Sensoren und der Heterogenität der Knoten. Folglich sollten Schätzalgorithmen robust gegenüber Paketverzögerungen und -ausfällen sein, die lokale Verarbeitung gemäß der globalen Sensornetzkapazität optimieren und vorhandene Rechen- und Speicherkapazitäten nutzen.

In der Literatur werden grundsätzlich drei Abstufungen hinsichtlich der Informationsverarbeitung in Sensornetzen unterschieden. In der klassischen Schätzung werden alle Informationen an einem zentralen Rechenknoten verarbeitet. In der verteilten Schätzung wird dieses Konzept auf eine Vorverarbeitung von Messungen in den Sensoren erweitert, wobei das Optimierungskriterium weiterhin die Schätzqualität an einem dedizierten Fusionsknoten ist. In der allgemeinsten Form wird ein Verbund von Knoten betrachtet, indem Schätzungen dezentral optimiert und mit anderen Knoten im Sensornetz ausgetauscht und fusioniert werden. Für lineare Systeme mit zentraler Verarbeitung ermöglicht das Kalman Filter eine gemäß mehrerer Kriterien (Maximum-Likelihood, minimale quadratische Abweichung, etc.) optimale Schätzung. Die Verallgemeinerung dieses Verfahrens von zentralen auf verteilte und dezentrale Systeme mit informationsbeschränkten Sensoren bildet den Kern dieser Arbeit.

Die wesentlichen Herausforderungen bei der Generalisierung der linearen Verarbeitung auf verteilte Systeme bestehen in der lokalen Filterung, also der rekursiven Kombination von Messungen in den Sensoren, und der Fusion von Schätzungen. Zur effizienten Lösung der letztgenannten Herausforderung werden Methoden zur Quantifizierung und Integration von unpräzise bekannten Abhängigkeitsstrukturen in Sensornetzen entwickelt. Dies ist nötig, da nur unter Berücksichtigung von Abhängigkeiten in der Fusion eine Unterscheidung zwischen exklusiven und bereits verfügbaren Informationen möglich ist. Während sich Arbeiten in der Literatur auf das konservative Abschätzen von unbekannten Korrelationen beschränken, werden im

Zuge der Arbeit zwei Ansätze entwickelt, die, für die betrachteten linearen Systeme, basierend auf der Zerlegung von Fehlergrößen in Rauschanteile die Rekonstruktion von Abhängigkeiten anhand von lokal errechneter Variablen erlauben. Die Präzision der Rekonstruktion kann in beiden Verfahren relativ zu steigendem Rechen- und Kommunikationsaufwand skaliert werden und ermöglicht somit die Implementierung von zugeschnittenen Lösungen für beliebige Sensornetze.

Trotz begrenzten Wissens über die Verarbeitung anderer Sensoren muss die Filterung gemäß Eigenschaften des Sensornetzes optimiert werden, um eine bestmögliche Ausbeute der Informationen in Messungen zu erreichen. Ein in der Disseration entwickelter Algorithmus benutzt zu diesem Zweck eine Hypothese über die Messqualität des gesamten Sensornetzes und stellt Techniken zur robusten Korrektur von unzutreffenden Hypothesen bereit. Es wird nicht nur gezeigt, dass bei zutreffenden Annahmen der Algorithmus trotz verteilter Verarbeitung die gleichen (optimalen) Resultate wie ein zentrales Kalman Filter erzielt, sondern darüber hinaus theoretisch und in Evaluationen nachgewiesen, dass bessere Ergebnisse als mit parallelen Kalman Filtern und anschließender Fusion erreicht werden. Damit empfiehlt sich die Implementierung der zugrundeliegenden Idee nicht nur in klassischen Sensornetzwerkszenarien, sondern zum Beispiel auch zur schritthaltenden Selbstlokalisierung, die typischerweise auf die lokale Vorverarbeitung von Gyrometer-, Beschleunigungssensor- und Schrittmesserdaten angewiesen ist.

Die Verknüpfung der entwickelten Techniken erlaubt die Optimierung der stochastischen Verarbeitung in Sensornetzen in einer Vielzahl von Anwendungsfällen. Da die hergeleiteten Methoden für Spezialfälle optimale Ergebnisse liefern und die Kombination der Verfahren anhand theoretischer Einsichten möglich ist, trägt die Arbeit wichtige Erkenntnisse zur dezentralen Informationsverarbeitung bei. Diese Erkenntnisse können insbesondere zu einer effizienteren Ausnutzung von Sensorressourcen in einer Vielzahl praktischer Probleme führen.

Abstract

The integration of information and assumptions from different sources that bear the potential to deduce valuable conclusions appeals to researchers and engineers equally and constitutes a fascinating field of study. There are compelling reasons to argue that the associated theory is becoming even more and more important in an increasingly interconnected world. Under the topic of stochastic information processing, the mathematical foundation of the statistical evaluation and interpretation of uncertain data is laid. Estimation theory in sensor networks constitutes an important part that focuses on the distributed processing of information subject to locally constrained knowledge.

While, originally, sensor networks were perceived as collections of miniature systems that jointly monitor spatially distributed phenomena subject to weak computing and limited battery capacities, it has become evident in recent years that the concept and the associated ideas are applicable to various technologies. The incorporation of data from different sensor devices in smartphones and the distributed calculation of complex weather simulations constitute only two examples of application areas. The ever-increasing computational demand and the concomitant growing complexity requires even in traditionally centralized systems the decomposition of operations in functionally independent subsystems with requirements that correspond to those of traditional sensor network nodes.

This thesis deals with sensor network algorithms for the stochastic estimation of a state that is only accessible via uncertain observations. This problem is encountered in several areas in literature and is discussed under the terms target tracking, distributed and decentralized estimation, and multisensor data fusion. The additional challenges of a typical sensor network compared to the central processing of information manifest themselves in the uncertain (often wireless) communication, limited information at local sensors, and the heterogeneity of the nodes. Thus, proper estimation algorithms must cope with packet delays and losses, optimize the processing based on local knowledge, and take advantage of available computing and storage capacities.

Basically, a distinction between three schemes for the processing of information in sensor networks is made in literature. In classical estimation theory, all observations are processed at a central computing node. Distributed estimation considers the preprocessing of measurements at the sensors with the objective to optimize the estimate at a dedicated fusion node. In the most general form, decentralized nodes cooperatively enhance their local estimates by exchanging and merging information with their neighborhood. For centralized linear systems with Gaussian distributed random variables, the Kalman filter constitutes the optimal estimator subject to several criteria (maximum likelihood, mean squared error, etc.). This thesis now deals with the generalization of Kalman filter ideas to distributed and decentralized systems with information-constrained sensors.

The main challenges arising from the generalization of the centralized processing are the filtering at the sensors, i.e., the recursive combination of measurements, and the fusion of estimates. The latter challenge is addressed with methods for the quantification and integration of imprecise dependency information. The exploitation of dependencies is necessary to allow an identification of mutually exclusive information in the fusion. While previous work in literature is confined to the bounding of unknown correlations, two approaches for the reconstruction of dependencies are developed in this thesis. Both are based on a decomposition of error quantities into individual noise terms and can be implemented at the sensors without knowledge about remote processing. The precision of both techniques is adjustable subject to increasing computational and communication effort and thus, permits deriving tailored solutions for various sensor network setups.

In order to overcome the first challenge and to realize an efficient filter processing at the sensors, the combination of observations is optimized according to a hypothesis about the measurement capacity of the sensor network. As assumed parameters typically do not accurately reflect the reality, techniques for robust correction of incorrect hypotheses are provided. It is not only shown that this algorithm achieves the same (optimal) results as a central Kalman filter for suitable hypothesis but moreover, it is demonstrated in theory and evaluations that the proposed procedure provides preciser estimates than standard approaches that use local Kalman filters. Thus, the implementation of approaches that operate on hypotheses is beneficial in various distributed estimation scenarios such as for example for self-localization, which typically relies on a preprocessing of accelerometer, gyro, and pedometer data.

A combination of the developed techniques allows optimized stochastic processing in a variety of applications. Since the proposed methods provide optimal results for special cases, e.g., distributed estimation, and the combination of them on the basis of theoretical insights is viable, the insights are directly applicable to decentralized information processing and can contribute to a more efficient exploitation of sensor resources in a wide range of practical problems.

CHAPTER 1

Introduction

The perception and processing of information is ubiquitous in human and technical reasoning. It is an inherent attribute of real world problems that desired information is rarely directly accessible but instead the result of an implicit or explicit evaluation of observations or related facts. As a consequence, the systematic analysis of information provides no guaranteed statements but approximations that are uncertain by nature. Indeed, without proper assessment, information is difficult to process and the combination of multiple uncertain sources may even reduce its quality. In particular, a bias can be induced in the fusion result, and the quality of combined information is simply overestimated. Due to increasing awareness of these problems and due to advances in different areas of research, explicit treatment of uncertainty has found its way in various applications. Even though stochastic inference can necessitate sophisticated modeling, the resulting predictions and decisions are not only comprehensible from a mathematical point of view but also reliable in practical use.

In an increasingly interconnected world, uncertain observations stem from different, often heterogeneous sources. It then remains concealed to the receiver, what kind of prior processing the sources performed on the observations and whether information is a mutual dependent copy of already available data. Smartphones serve as a perfect example of such modern sensor networks. As a matter of fact, these devices are conglomerations of various sensors such as microphones, GPS modules, video cameras, etc. [105]. Already in the devices, data from different sensors is merged to offer enhanced

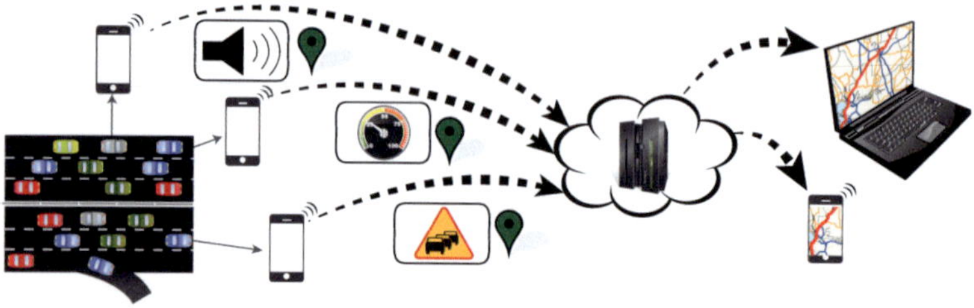

Figure 1.1: Several smartphones observe independently of each other (uncertain) data such as honking levels, speed, or user inputs. This information is locally preprocessed, enriched with uncertain GPS data, and sent to a server. There, the received packages are combined to gain new information, e.g., about traffic jams.

services such as videotelephony and augmented reality [164]. Even more important, the interconnection of smartphones permits the cost-efficient integration of spatiotemporally distributed information as it is discussed in literature under the catchphrase "mobile crowd sensing" [50]. Examples for such services are the detection of traffic jams as depicted in Figure 1.1 and the systematic storage of pictures marked with GPS data that allows monitoring temporal changes at different spots on earth [111]. There are compelling reasons to argue that these ideas are only the beginning, and more and more services will rely and exploit loosely related information in the future.

The integration of information from heterogeneous sensors that operate independently of each other confronts scientists and practitioners with new problems not encountered in classical sensor network theory, which has been developed mainly in the context of well-controlled sensor environments such as indoor localization [71], hazard detection [72], and alike. Modern estimation algorithms must process information with unknown origin and dependencies. In particular, quite often not even the number of involved sensors can be anticipated. In such scenarios, a homogeneous optimization of sensors is hardly realizable. Additionally, even if technical advances have relieved some restrictions, the challenges from classical sensor network theory have not yet outlived themselves. These include, primarily, limited battery capacities at the sensors that manifest themselves in a constrained operationality with respect to energy-intensive computational and communication tasks. Moreover, quite often a wireless communication is established that is inherently linked to the risk of packet delays and losses.

Certainly, information-processing systems must feature high degrees of adaptability and flexibility to prove effective in such environments. This pertains to the local processing of observations that can hardly rely on sensor network characteristics, which are quite often unknown at the time of sensing and certainly concealed to the sensor, but also to the fusion of information, which should permit a consideration of dependencies without requiring it as a condition.

1.1 Problems in Sensor Network Estimation

In order to systematically deal with the aforementioned difficulties, information processing is embedded in a mathematical framework where the value of interest is assumed to be an unknown state vector and observations are stochastically distributed according to a known relation to the state. The challenge is then to compose **estimates** from observations such that the difference to the underlying state is minimized. In this abstract analysis, primarily, two operations need to be considered. For one thing, the combination of estimates and measurements is examined under the topic filtering and focuses on the integration of information that quite often pertains only to a small part of the state. In contrast to that, the term fusion describes the combination of two or more uncertain estimates.

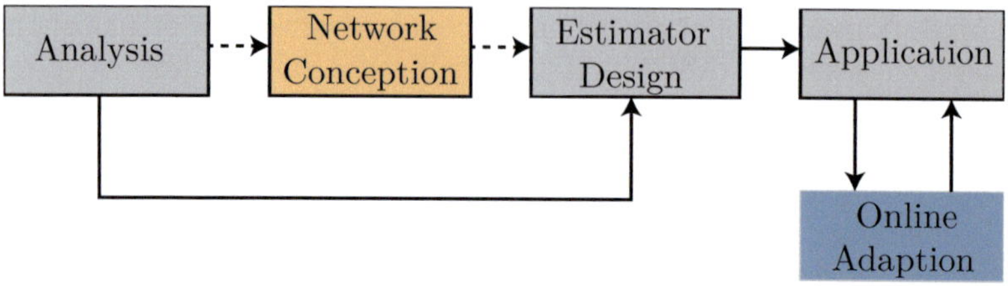

Figure 1.2: Different processes for the design of sensor network estimators. In classical sensor networks, the concept is known when the estimator is derived and only small adaptions are necessary once the network is deployed. In the more general interpretation of sensor networks considered in this thesis, the network structure can change completely during runtime, which necessitates a high degree of flexibility of the estimators.

It follows not only from a mathematical analysis but also from intuition that the optimal **fusion** depends on the quality of individual estimates and their dependencies. As a matter of fact, dependencies emerge between estimates because the same stochastic system is observed and due to data exchanges between nodes that lead to increasingly similar information at the sensors. As the processing of remote sensors is barely assessable and the history of past data exchanges hardly traceable, accurate knowledge about these dependencies is often not given. Indeed, for the subsequent use of estimates in decision-making or control systems, implementations of the fusion operation should not only minimize the error of the estimate but also provide a quality assessment of the fused estimate. Thus, the questions how to quantify dependencies subject to knowledge constraints of individual sensors and how to fuse estimates subject to imprecise quality and dependency information arise.

The preprocessing of measurements and their recursive aggregation at the sensors are key components of every estimator and determine to a large part the quality of estimates in sensor networks. The problem that needs to be solved in recursive **filtering** is to determine the function by which measurements are to be combined into a local estimate that is subsequently transmitted to other nodes. This is, in particular, important as the optimal estimate cannot be recovered from locally preprocessed values when the presence of the sensor network is neglected in the optimization of local filters, e.g., by applying Kalman filters at the sensors. In summary, the challenge is to find optimized filter matrices at the nodes subject to imprecisely known sensor network properties.

Concurrent filtering and fusion at several nodes, furthermore, leads to interdependencies that renders sensor network estimation a difficult scientific challenge with many open problems. In particular, the considered estimation problems exhibit many degrees of freedom so that practical implementations and theoretical properties are only achieved in special cases or when additional assumptions are imposed.

1.2 Outline and Overview

Hence, this thesis focuses on **linear processing** of uncertain information and the mean squared error loss function. This is a well-established framework and has been the basis for key results in sensor network estimation. In contrast to the problem setting of classical distributed estimation, the challenges imposed by modern sensor networks are also addressed by considering limited information at the sensors. Instead of concurrently optimizing filter and fusion processing of the entire sensor network prior to application, sensors are supposed to operate on imprecise and incomplete information about the sensor network and to adapt online to changes in the network structure as illustrated in Figure 1.2. Essentially, the objective is to derive estimation algorithms that handle not only high uncertainty in the data but also highly uncertain and unforeseeable models.

Chapter 2: Fundamentals of Estimation Theory serves as an introduction and mathematical classification of the estimation problem. The process of extracting models from real world phenomena is described, and fundamental results of estimation theory in the Bayesian framework are presented. The focus is laid on challenges in sensor network estimation. For this purpose, three types of sensor networks are delineated that reflect increasingly complex problem settings. The distribution of calculations away from a central node to the sensors introduces sources of dependency between estimates. For example, the evolution of the underlying state is accompanied by uncertainties that pertain to all estimates equally. Additionally, due to the exchange of information in the sensor network, estimates increasingly incorporate the same information. These dependencies interrelate estimates and demand the concurrent optimization of filter and fusion processing at distributed sensors. State-of-the-art methods are illuminated, and their applicability to modern sensor networks with limited sensor knowledge is discussed.

Chapter 3: Information Fusion in Sensor Networks considers the merging of several estimates. In that chapter, two approaches are pursued to quantify dependencies in sensor networks. An estimation scheme based on samples is proposed that gives stochastic approximations of covariances. For an implementation in sensor networks, optimal parameters for finite samples are derived. The other technique draws on existing conservative bounding theory. Based on a decomposition of covariances, linear dependencies between estimates can be reconstructed partially even if sensors operate on local data only. Implementations of this idea are presented, and generalizations of the bounding theory in terms of the number of considered estimates and the incorporation of dependency information are derived. Moreover, fundamental properties are obtained that apply to the underlying bounding techniques and fill existing gaps in literature. The proposed algorithms bridge the gap between well-known techniques for the fusion under either known or unknown dependencies.

Chapter 4: Hypothesizing Distributed Kalman Filtering is concerned with the distributed optimization of filter operations in the presence of a dedicated fusion center. A key prerequisite for an optimal distributed estimation scheme is to ensure a relative weighting of local measurements that is equal to that of a centralized estimation scheme. Hence, techniques are proposed that use hypotheses to satisfy this constraint best possible for different degrees of knowledge about the sensor network structure. An important insight of the chapter is that the subsequent fusion of information demands adapted filter processing. This notion directly transfers to sensor network estimation without a dedicated fusion center and the processing of out-of-sequence measurements.

CHAPTER

2

Fundamentals of Estimation Theory

Observing and processing uncertain information about our environment is an integral part in decision-making of humans and machines. For example, human drivers implicitly approximate the distance to traffic lights based on natural intuition and experiences and lean on their **estimates** when they brake the car. A more explicit processing that illuminates estimation theory is positioning the car by means of navigation devices. A widespread concept for localization is to utilize satellite navigation such as GPS, which aims to determine positions from time differences in signals. Despite intensive efforts, an uncertainty in the position remains induced by unpredictable delays in the transmission and technical limitations in the signal processing [37]. Hence, it is an inherent characteristic of satellite-based localization methods that measurements are imprecise.

In a more general perspective, an **unknown state** is to be determined from one or multiple **observations** that permit the inference of some information about the unknown quantity. The objective of estimation theory boils down to deriving approximations of the state that satisfy meaningful quality properties. In literature, various research directions of estimation theory are pursued that often provide different perspectives on similar algorithms and results. In the early stages of estimation theory, the functional relation between observations and state was assumed known and parameters were optimized by means of **deterministic analyses**. In this context, residuals, i.e., deviations of the observations from the estimated model are minimized by means of methods such as least squares. Indeed, deterministic approaches accept and process residuals as outcomes of the approximation but cannot, taken in isolation, explain their emergence. This is achieved by deriving estimates premised on the assumption that measurements are outputs of stochastic experiments. **Stochastic models** allow the systematic inference of measurements based on probabilistic principles and the optimization of estimates according to well-established quality attributes from probability theory. However, probability theory itself knows different interpretations that are closely bound up with the question whether probability means a physical tendency that something happens or a belief that is occurs.

The first-mentioned interpretation is referred to as classical or **frequentist** statistics. Measurements are construed as occurrences of possible events of an experiment. The state is a fixed deterministic parameter of the stochastic model and estimates are to maximize a utility or to minimize a loss function. In the **Bayesian** perspective, estimates are seen as conditional probabilities that express confidence or a degree of knowledge about the state. This concept necessitates, by definition, prior knowledge about the unknown state. Indeed, the parameters are random variables. Consequently, the estimate is random itself. As a matter of fact, the admissibility of different interpretations of probability has fostered a debate on the correct and best perspective, which is delineated in [13].

For the subsequent examinations, the Bayesian point of view is taken. Let $\underline{z}_1, \ldots, \underline{z}_m$, $m \in \mathbb{N}$ denote measurements that provide information about the true state $\underline{\mathbf{x}}$. Then, probabilistic inference with Bayes' theorem according to

$$p(\underline{\mathbf{x}}|\underline{\mathbf{z}}) = \frac{p(\underline{\mathbf{z}}|\underline{\mathbf{x}})p(\underline{\mathbf{x}})}{p(\underline{\mathbf{z}})} = \frac{p(\underline{\mathbf{z}}|\underline{\mathbf{x}})p(\underline{\mathbf{x}})}{\int_{\underline{x}\in\mathbb{R}^{n_x}} p(\underline{\mathbf{z}}|\underline{x})p(\underline{x})d\underline{x}} \tag{2.1}$$

permits integrating likelihoods $p(\underline{\mathbf{z}}|\underline{\mathbf{x}})$ into the prior distribution $p(\underline{\mathbf{x}})$ in order to obtain the posterior conditional density $p(\underline{\mathbf{x}}|\underline{z}_1, \ldots, \underline{z}_k)$

that reflects the knowledge about the state. In the probabilistic model, the evolution of the state can be described by means of the Chapman-Kolmogorov equation [83] such that the estimation process is completely characterized by the density representation. Even though Monte Carlo and deterministic **sampling methods** have been developed to approximate the evolution of these conditional densities, their calculation proves difficult for arbitrary densities and transformations. More details on this topic are given in Section 2.2.

A possibility to facilitate probabilistic inference is to pose the question what is eventually inferred from the conditional density. From simple comparisons of densities in decision theory to the calculation of confidence intervals, various properties can be extracted from random variables. In this thesis, the focus is laid on the derivation of **point estimators** that find application in many domains such as target tracking and control theory.

This chapter serves as the mathematical introduction to estimation and filter theory and provides a comprehensive overview of state-of-the-art methods relevant for subsequent considerations concerned with sensor network estimation. At the outset of every estimator design and of this thesis stands the question how to transform a real world problem into a suitable mathematical model. The systematic process and associated problems are examined in Section 2.1. As result of these considerations, a linear model is presented that constitutes a compromise between goodness of approximation quality and mathematical simplicity.

The basics of Bayesian estimation theory are laid in Section 2.2 by deriving the minimum mean squared error point estimator and its computationally attractive linear counterpart. The relation between the two estimators is examined and the Kalman filter is presented as the optimal solution for linear models. Moreover, alternative approaches for central estimation and the treatment of nonlinear models are examined and presented.

An introduction to estimation with several sensors follows in Section 2.3. Different parameters to describe the interaction between sensors are discussed and in an attempt to unify the sensor network theory, a classification in three types is proposed. A detailed analysis and state-of-the-art is provided for all three types. Finally, key research questions and associated challenges for estimation in modern sensor networks are identified and motivated in Section 2.4.

2.1 Modeling Real-world Problems

At the outset of conceptualizing an estimator for a given problem, underlying state and mathematical models have to be identified. In general, finding these quantities is nontrivial and has been extensively studied under the topic **system identification** [110]. As the focus of this thesis is on the processing and optimization of sensor network estimators rather than model identification, the following explanation is confined to the basic concepts. For a more comprehensive discussion of challenges and results related to system identification, the reader is referred to [96, 110, 155].

The process of deriving a model from the real world is as follows [83]: first, relevant variables that depict the process are identified and linked according to **causal and physical relations**. Alternatively, black-box models must be employed that operate exclusively on observed data and aim to find the best representative from a set of candidates according to a specified criterion. Then, the derived model is simulated, validated, and improved until no additional improvements are achieved. Usually, at this point, the model is still not a perfect reflection of the real world because the system behavior is partially unpredictable, the instruments' precision is limited, etc. A popular method to cope with these challenges is to introduce stochastic noise terms. Hence, at the end of a system identification process stands a **stochastic differential or difference equation**, which, then, is used to obtain proper estimators.

Fortunately, (simplified) causal and physical relations of real world variables have already been derived in various areas. Consider for example the movement of an object with known mass that is subject to a constant acceleration and let an observer measure the position of the object at several time instances. Then, the mathematical relation between position, velocity, and acceleration can be exploited to determine the force that is applied to accelerate the object by means of Newton's second law. If the obtained force insufficiently reflects the reality, the uncertainty in the movement, acceleration, and observation processes can be explicitly modeled by stochastic noise terms. A less obvious example stems from financial markets. Given the complexity of price formations and the number of interacting traders with contradictory aims, it is surprising that the price for some bets on the future is uniquely determined by observable market variables and an expected volatility. In fact, the value of certain derivatives of financial products is obtained by a partial differential equation, termed the Black–Scholes PDE [18, 115]. Further applications of system identification to real world problems can be found in areas such as biology, physics, and engineering [83].

2.1.1 State Space Models

The result of the system identification process constitutes a stochastic differential equation of a pertinent state variable as well as the functional relation between state and observations. Throughout the thesis, let the **state** be a minimal set of variables that define the system behavior according to the following two criteria [128]:

1. **State Evolution Property**: Initial state and inputs over the considered time period determine the state.

2. **Instantaneous Output Property**: Outputs, i.e., measurements, are given as a function of state and inputs.

For the sake of simplicity, differential equations are assumed given as first order systems. Then, random processes are Markovian. Therefore, all information relevant for the future is entailed in the state of the latest time step. Precise models of real world phenomena are time-continuous, leading to stochastic differential equations (SDE)

$$\frac{d\underline{\mathbf{x}}_t}{dt} = \underline{f}(\underline{\mathbf{x}}_t, \underline{\mathbf{w}}_t, t) \ , t \geq 0 \ , \tag{2.2}$$

with unknown random state variables $\underline{\mathbf{x}}_t \in \mathbb{R}^{n_x}$. As it is hardly feasible to derive efficient estimators for the general formulation (2.2), special models that either precisely describe or approximate the considered system are of particular interest. Motivated by the central limit theorem, it is reasonable to confine the attention to stochastic systems with additive white Gaussian noise $\underline{\mathbf{w}}_t$ that is linearly transformed with matrices $\mathbf{G}(\underline{\mathbf{x}}_t, t)$, which boils down to the well-known Langevin equation

$$\frac{d\underline{\mathbf{x}}_t}{dt} = \underline{f}(\underline{\mathbf{x}}_t, t) + \mathbf{G}(\underline{\mathbf{x}}_t, t)\underline{\mathbf{w}}_t \ . \tag{2.3}$$

For SDEs in the form (2.3), a comprehensive mathematical framework has been developed. When $\mathbf{G}(t)$ is deterministic, the theory around the Wiener filter [169] is applicable. More general techniques for stochastic $\mathbf{G}(\underline{\mathbf{x}}_t, t)$ have been devised by means of the Itô stochastic calculus [81] and the Stratonovich integral [159]. For special cases, such as $\underline{f}$ linear and independent of the state and for observations at discrete time steps, the efficient estimator, i.e., the continuous time Kalman filter [93], can even be given in closed form.

However, in modern digital systems, variables are often processed at **discrete time steps**, which has led to extensive research of the discrete-time counterparts to (2.2) and (2.3). If discrete-time models are not constructed directly from observations, they must be calculated from the corresponding SDEs by integration. For arbitrary SDEs, this process implies an approximation. For example, the derivation of the discrete-time motion process of a particle depends on how the acceleration is modeled in the discretized time [106].

The general form of **stochastic difference equations** is given by the following nonlinear equation

$$\underline{\mathbf{x}}_{k+1} = \underline{f}(\underline{\mathbf{x}}_k, \underline{u}_k, \underline{\mathbf{w}}_k) \ , \ \ k \in \mathbb{N}_0 \ . \tag{2.4}$$

Again, it is worthwhile to consider models with additive noise, leading to the simpler form

$$\underline{\mathbf{x}}_{k+1} = \underline{f}(\underline{\mathbf{x}}_k, \underline{u}_k) + \underline{\mathbf{w}}_k \ . \tag{2.5}$$

Indeed, without observations about the unknown state, the randomness in the state model renders estimates imprecise over time. Therefore, it is assumed that measurements are observed at discrete time steps and are related to the true state by a known function. The general form is given by

$$\underline{\mathbf{z}}_k = \underline{h}(\underline{\mathbf{x}}_k, \underline{\mathbf{v}}_k) \ , \tag{2.6}$$

with measurements $\underline{\mathbf{z}}_k \in \mathbb{R}^{n_z}$, where models

$$\underline{\mathbf{z}}_k = \underline{h}(\underline{\mathbf{x}}_k) + \underline{\mathbf{v}}_k \ , \tag{2.7}$$

with additive noise are of particular interest. Among the models with additive noise (2.5) and (2.7), especially the ones with linear transition and measurement functions have attracted attention as they permit deriving computationally efficient estimators with useful theoretical properties.

Definition 2.1 *In* ***linear systems*** *(with uncorrelated noise terms), the state evolves according to the model*

$$\underline{\mathbf{x}}_{k+1} = \mathbf{A}_k \underline{\mathbf{x}}_k + \underline{\mathbf{w}}_k \ \textit{with} \ \mathrm{E}\{\underline{\mathbf{w}}_k\} = \underline{0} \ , \mathrm{E}\{\underline{\mathbf{w}}_k (\underline{\mathbf{w}}_k)^\top\} = \mathbf{Q}_k \ , \tag{2.8}$$

and measurements are observed according to

$$\underline{\mathbf{z}}_k = \mathbf{H}_k \underline{\mathbf{x}}_k + \underline{\mathbf{v}}_k \ \textit{with} \ \mathrm{E}\{\underline{\mathbf{v}}_k\} = \underline{0} \ , \mathrm{E}\{\underline{\mathbf{v}}_k (\underline{\mathbf{v}}_k)^\top\} = \mathbf{R}_k \ . \tag{2.9}$$

Different noise terms $\{\underline{\mathbf{w}}\}_{k \in \mathbb{N}_0}$ *and* $\{\underline{\mathbf{v}}\}_{k \in \mathbb{N}}$ *are assumed to be uncorrelated from each other.*

Linear transformations $\mathbf{G}$ of zero-mean noise terms $\tilde{\underline{\mathbf{w}}}_k$ yield a noise term $\underline{\mathbf{w}}_k$ with $\mathrm{E}\{\underline{\mathbf{w}}_k\} = \underline{0}$ and $\mathrm{E}\{\underline{\mathbf{w}}_k(\underline{\mathbf{w}}_k)^\top\} = \mathbf{G}\mathbf{Q}_k(\mathbf{G})^\top$, and, consequently, are covered in Definition 2.1. Deterministic inputs $\underline{u}_k$, which are relevant for control applications, can be considered in the same framework with the system model

$$\underline{\mathbf{x}}_k = \mathbf{A}_k \underline{\mathbf{x}}_k + \mathbf{B}_k \underline{u}_k + \underline{\mathbf{w}}_k \ . \tag{2.10}$$

In summary, linear systems constitute a tradeoff between practical applicability and theoretical rigor. They have become popular with the Kalman filter [93] and are obtained from real world setups directly or are derived by means of linearization techniques, as it will be discussed in Section 2.2.2. Especially for Gaussian distributed noise, the linear model has been intensively researched in estimation and control literature and serves as the basic model for the examinations in this thesis.

Eventually, it is worth formalizing the statement that estimates become useless with passing time when no measurements are observed. In fact, a rigorous examination reveals that the difference between estimate and true state should not grow unboundedly.

Definition 2.2 *A system with initial state $\underline{\mathbf{x}}_0$ is said to be* ***observable****, if and only if the initial state can be obtained from the system's outputs.*

When observability is achieved for all feasible initial values, the system is called **completely observable**. For time-invariant linear systems, i.e., if the matrices in (2.8) and (2.9) remain constant over time, observability can be checked by means of the observability matrix

$$\mathcal{O} = \begin{pmatrix} \mathbf{H} \\ \mathbf{H}\mathbf{A} \\ \mathbf{H}\mathbf{A}^2 \\ \vdots \\ \mathbf{H}\mathbf{A}^{n_x - 1} \end{pmatrix} \ . \tag{2.11}$$

If $\mathcal{O}$ has rank n_x, the system is observable. A slightly weaker condition with is **detectability**, which is satisfied when the non-observable states go to zero asymptotically. It is worth mentioning that there exist analogous concepts to observability and detectability in control literature called controllability and stabilizability.

2.2 Bayesian Point Estimators

In the Bayesian framework, a belief about the state is expressed by means of a conditional density that gives probabilities for all possible values. However, consider for example a hunter that aims to shoot a bird. Then, a **point estimate** of the bird's position is needed that maximizes the chance of a hit. Indeed, this common problem occurs also in control applications where a specific input has to be chosen. In order to extract a point estimate from the probabilistic description of the state, the expectation of a **loss or risk function** ℓ is to be minimized, i.e.,

$$\arg\min_{\underline{\hat{\mathbf{x}}}(\underline{\mathbf{z}})} \left[\mathrm{E}_{\underline{\mathbf{x}},\underline{\mathbf{z}}}\{\ell(\underline{\hat{\mathbf{x}}}(\underline{\mathbf{z}}), \underline{\mathbf{x}})\} \right] \tag{2.12}$$

needs to be solved for norms such as ℓ_1, ℓ_2, ℓ_∞, or combinations of them [119]. Popular in literature are the ℓ_∞ norm [153], which aims to provide robust worst-case estimates, and the MSE loss ℓ_2

$$\arg\min_{\underline{\hat{\mathbf{x}}}(\underline{\mathbf{z}})} \left[\mathrm{E}_{\underline{\mathbf{x}},\underline{\mathbf{z}}} \left\{ \sqrt{(\underline{\hat{\mathbf{x}}}(\underline{\mathbf{z}}) - \underline{\mathbf{x}})^\top (\underline{\hat{\mathbf{x}}}(\underline{\mathbf{z}}) - \underline{\mathbf{x}})} \right\} \right] \tag{2.13}$$

that serves as optimization criterion in this thesis. The estimator $\underline{\hat{\mathbf{x}}}(\underline{\mathbf{z}})$ that minimizes (2.13) is called minimum MSE (MMSE) estimator. The following description of the fundamental results of the MSE theory is loosely based on [47, 128, 131].

In a first step, consider a scalar state $\mathbf{x}$ and a vector of measurements $\underline{z}$. Then, a problem that is closely related to (2.13) is the minimization of the ℓ_2 norm given a specific measurement, i.e., the derivation of $\hat{\mathbf{x}}(\underline{z})$ that minimizes

$$\begin{aligned} \mathrm{E}_{\mathbf{x}|\underline{\mathbf{z}}}\{\ell_2(\hat{\mathbf{x}}(\underline{z}),\mathbf{x})|\underline{\mathbf{z}}=\underline{z}\} &= \mathrm{E}_{\mathbf{x}|\underline{\mathbf{z}}}\{(\hat{\mathbf{x}}(\underline{z})-\mathbf{x})^2|\underline{\mathbf{z}}=\underline{z}\} \\ &= \int_{x\in\mathbb{R}^{n_x}} (\hat{\mathbf{x}}(\underline{z})-x)^2 p(x|\underline{z})dx \ . \end{aligned} \tag{2.14}$$

By equating the derivative

$$\frac{d\,\mathrm{E}_{\mathbf{x}|\underline{\mathbf{z}}}\{\ell_2(\hat{\mathbf{x}}(\underline{z}),\mathbf{x})|\underline{\mathbf{z}}=\underline{z}\}}{d\hat{\mathbf{x}}(\underline{z})} = 2\cdot\int_{x\in\mathcal{X}} (\hat{\mathbf{x}}(\underline{z})-x)p(x|\underline{z})dx$$

to zero, an extreme is found to be

$$\hat{\mathbf{x}}(\underline{z}) = \int_{x\in\mathbb{R}^{n_x}} x\cdot p(x|\underline{z})dx = \mathrm{E}_{\mathbf{x}|\underline{\mathbf{z}}}\{\mathbf{x}|\underline{\mathbf{z}}=\underline{z}\} \ .$$

The second derivative is 2. Thus, for a given measurement, the optimal point estimate is the conditional expectation $\mathrm{E}_{\mathbf{x}|\underline{\mathbf{z}}}\{\mathbf{x}|\underline{\mathbf{z}}=\underline{z}\}$.

The MSE criterion (2.13) is defined based on the expected value of the random measurement vector and not for individual measurements. However, no constraints have been imposed on the measurement vector to derive the solution of (2.14), and so, the optimal estimator in the MMSE sense is the rule that obtains the conditional expectation for each measurement. Mathematically speaking, measurements are modeled as random vector. It follows from $p(\underline{\mathbf{x}},\underline{\mathbf{z}}) = p(\underline{\mathbf{x}}|\underline{\mathbf{z}})p(\underline{\mathbf{z}})$ that

$$\mathrm{E}_{\mathbf{x},\underline{\mathbf{z}}}\{(\hat{\mathbf{x}}(\underline{\mathbf{z}})-\mathbf{x})^2\} = \int_{\underline{z}\in\mathbb{R}^{n_z}} \underbrace{\mathrm{E}_{\mathbf{x}|\underline{\mathbf{z}}}\{(\hat{\mathbf{x}}(\underline{z})-\mathbf{x})^2|\underline{\mathbf{z}}=\underline{z}\}}_{\text{minimal}}\underbrace{p(\underline{z})}_{\geq 0}\,d\underline{z} \ ,$$

where the inner term is optimally solved by the conditional expectation $\mathrm{E}_{\mathbf{x}|\underline{\mathbf{z}}}\{\mathbf{x}|\underline{\mathbf{z}}=\underline{z}\}$. In the sequel, let the expectation $\mathrm{E}\{\cdot\}$ be taken over all random vectors, if not otherwise stated. As proven in the appendix by means of the **Orthogonality Principle** (Theorem A.1), the scalar MMSE estimator generalizes to multivariate systems according to the following theorem.

Theorem 2.1 *The conditional expectation* $\mathrm{E}\{\underline{\mathbf{x}}|\underline{\mathbf{z}} = \underline{z}\}$ *minimizes the MSE from* (2.13).

PROOF. See Appendix A. □

Hence, the MMSE estimator is the expectation of the state density conditioned on observed measurements. While it is advantageous that the MMSE estimator is determined by a simple relation to the measurements, it is unfortunate that the conditional density needs to be propagated and assessed.

Popular methods to approximate the desired conditional density are sequential Monte Carlo methods [40, 74], which are referred to as **particle filters** in recursive estimation theory [64, 147, 152]. However, these methods operate on stochastic sample representations of densities that tend to provide a sparse and imprecise coverage for high-dimensional state spaces, i.e., they suffer from the "curse of dimensionality" [14]. This is aggravated by the fact that for nonlinear system transformations, distributions are in general not conjugated [47], i.e., posterior distributions are from other families than prior distributions. In particular, even if the prior is Gaussian distributed and can be represented by mean and covariance, transformations in nonlinear systems (2.4) and (2.6) yield distributions from different families that can be multi-modal and can require representations with infinitely many variables.

Apart from Monte Carlo approaches, several techniques are discussed in literature to cope with the propagation of densities. When system and measurement equations permit the calculation of moments of the posterior density in closed form, the MMSE estimator can be obtained directly. Otherwise, **deterministic sampling methods** can be employed that allow covering the probability space more efficiently than Monte Carlo approaches. The idea is to derive (Dirac or Gaussian) mixture approximations that minimize some distance measure to the true density [66, 69] and to propagate the mixture densities through the nonlinear functions. Indeed, for general densities, the minimization of appropriate distance measures involves considerable computational effort that either leads to long runtimes or requires to use pre-calculated sample set approximations [70, 157].

2.2.1 Linear Mean Squared Error Estimator

A computationally efficient approach is to confine the space of permissible estimators to linear functions of the form $\hat{\underline{\mathbf{x}}}(\underline{\mathbf{z}}) = \underline{\mathbf{m}} + \mathbf{K}\underline{\mathbf{z}}$. Then, the challenge is to find matrix $\mathbf{K}$ and vector $\underline{\mathbf{m}}$ that satisfy

$$\arg\min_{\hat{\underline{\mathbf{x}}}(\underline{\mathbf{z}})} \mathrm{E}\{\ell_2(\hat{\underline{\mathbf{x}}}(\underline{\mathbf{z}}) - \underline{\mathbf{x}})\} \text{ with } \hat{\underline{\mathbf{x}}}(\underline{\mathbf{z}}) = \underline{\mathbf{m}} + \mathbf{K}\underline{\mathbf{z}} \ , \tag{2.15}$$

i.e., to derive a linear minimum mean squared error (LMMSE) estimator. For the sake of a simple notation, the explicit dependency of the estimate on the measurement process is omitted in the following.

Note that $\mathbf{K}$ and $\underline{\mathbf{m}}$ do not depend on specific measurements $\underline{z}$ but only on the random measurement vector $\underline{\mathbf{z}}$. Consequently, LMMSE estimators minimize the expected error, i.e., the average error weighted with the probability of the occurrence of measurements. In particular, LMMSE estimators are in general suboptimal for individual measurement vectors.

In the sequel, the linear combination that minimizes the **MSE matrix** $\mathrm{E}\{(\hat{\underline{\mathbf{x}}} - \underline{\mathbf{x}})(\hat{\underline{\mathbf{x}}} - \underline{\mathbf{x}})^\top\}$ in the **positive definite sense** is derived. The positive definite relation establishes a partial ordering on symmetric matrices by claiming

$$\tilde{\mathbf{C}} \geq \mathbf{C} \Leftrightarrow \tilde{\mathbf{C}} - \mathbf{C} \geq \mathbf{0} \ , \tag{2.16}$$

where $\mathbf{C} \geq \mathbf{0}$, if and only if all eigenvalues of $\mathbf{C}$ are larger than or equal to 0 [76]. Note that $\tilde{\mathbf{C}} \geq \mathbf{C}$ implies that the diagonal elements of $\tilde{\mathbf{C}}$ are larger than the ones of $\mathbf{C}$. Hence, the minimization of the

MSE matrix in the positive definite sense minimizes, in particular, the trace of the MSE matrix, which, in turn, is the MSE. In a first step, the vector $\underline{\mathbf{m}}$ of the LMMSE estimator is derived.

Lemma 2.2 *Let* $\tilde{\underline{\mathbf{x}}}$ *denote a biased linear estimator with* $\mathrm{E}\{\tilde{\underline{\mathbf{x}}}\} \neq \mathrm{E}\{\underline{\mathbf{x}}\}$. *Then, the unbiased counterpart* $\hat{\underline{\mathbf{x}}} = \tilde{\underline{\mathbf{x}}} - \mathrm{E}\{\tilde{\underline{\mathbf{x}}} - \underline{\mathbf{x}}\}$ *has a smaller MSE matrix, i.e.,* $\mathrm{E}\{(\hat{\underline{\mathbf{x}}} - \underline{\mathbf{x}})^2\} \leq \mathrm{E}\{(\tilde{\underline{\mathbf{x}}} - \underline{\mathbf{x}})^2\}$.

PROOF. With $\mathrm{E}\{(\tilde{\underline{\mathbf{x}}} - \underline{\mathbf{x}})(\mathrm{E}\{\tilde{\underline{\mathbf{x}}} - \underline{\mathbf{x}}\})^\top\} = \mathrm{E}\{(\tilde{\underline{\mathbf{x}}} - \underline{\mathbf{x}})\}^2$, it holds

$$\begin{aligned}
\mathrm{E}\{(\hat{\underline{\mathbf{x}}} - \underline{\mathbf{x}})^2\} &= \mathrm{E}\{(\tilde{\underline{\mathbf{x}}} - \underline{\mathbf{x}} - \mathrm{E}\{\tilde{\underline{\mathbf{x}}} - \underline{\mathbf{x}}\})^2\} \\
&= \mathrm{E}\{(\tilde{\underline{\mathbf{x}}} - \underline{\mathbf{x}})^2\} - 2\,\mathrm{E}\{(\tilde{\underline{\mathbf{x}}} - \underline{\mathbf{x}})\}^2 + \mathrm{E}\{(\tilde{\underline{\mathbf{x}}} - \underline{\mathbf{x}})^2\}^2 \\
&\leq \mathrm{E}\{(\tilde{\underline{\mathbf{x}}} - \underline{\mathbf{x}})^2\}
\end{aligned}$$

for positive definite $\mathrm{E}\{(\tilde{\underline{\mathbf{x}}} - \underline{\mathbf{x}})\}^2$. □

It is worth pointing out the general validity of Lemma 2.2: linear biased estimators are improved in terms of the MSE by removing bias terms. Hence, the LMMSE estimator must be unbiased, i.e., the vector $\underline{\mathbf{m}}$ of the LMMSE estimator (2.15) must satisfy

$$\underline{\mathbf{m}} = \mathrm{E}\{\underline{\mathbf{x}}\} - \mathbf{K}\,\mathrm{E}\{\underline{\mathbf{z}}\} \ .$$

It remains to optimize the matrix $\mathbf{K}$. Let

$$\mathbf{C}^z = \mathrm{E}\{(\underline{\mathbf{z}} - \mathrm{E}\{\underline{\mathbf{z}}\})^2\}$$

denote the covariance of the random measurement vector and let

$$\mathbf{C}^{zx} = \mathrm{E}\{(\underline{\mathbf{z}} - \mathrm{E}\{\underline{\mathbf{z}}\})(\underline{\mathbf{x}} - \mathrm{E}\{\underline{\mathbf{x}}\})^\top\}$$

denote the cross-covariance matrix between state and measurement vectors. Then, the LMMSE estimator is given as follows.

Lemma 2.3 *The LMMSE estimator in the positive definite sense is given by*

$$\underline{\hat{\mathbf{x}}} = \mathrm{E}\{\underline{\mathbf{x}}\} + \mathbf{C}^{xz}(\mathbf{C}^{z})^{-1}(\underline{\mathbf{z}} - \mathrm{E}\{\underline{\mathbf{z}}\}) \ , \tag{2.17}$$

with

$$\mathbf{C} = \mathrm{E}\{(\underline{\hat{\mathbf{x}}} - \underline{\mathbf{x}})^2\} = \mathbf{C}^{x} - \mathbf{C}^{xz}(\mathbf{C}^{z})^{-1}\mathbf{C}^{zx} \ . \tag{2.18}$$

PROOF. Let the state covariance be denoted as

$$\mathbf{C}^{x} = \mathrm{E}\{(\underline{\mathbf{x}} - \mathrm{E}\{\underline{\mathbf{x}}\})^2\} \ .$$

Then, the MSE matrix of unbiased linear estimators is given by

$$\begin{aligned}
\mathrm{E}\{(\underline{\hat{\mathbf{x}}}(\underline{\mathbf{z}}) - \underline{\mathbf{x}})^2\} =& \mathrm{E}\{(\mathrm{E}\{\underline{\mathbf{x}}\} - \underline{\mathbf{x}} + \mathbf{K}(\underline{\mathbf{z}} - \mathrm{E}\{\underline{\mathbf{z}}\}))^2\} \\
=& \mathbf{C}^{x} + \mathbf{K}\mathbf{C}^{z}(\mathbf{K})^\top + \mathbf{K}\mathbf{C}^{zx} + \mathbf{C}^{xz}(\mathbf{K})^\top \\
=& (\mathbf{K} - \mathbf{C}^{xz}(\mathbf{C}^{z})^{-1})\mathbf{C}^{z}(\mathbf{K} - \mathbf{C}^{xz}(\mathbf{C}^{z})^{-1})^\top + \\
& \mathbf{C}^{x} - \mathbf{C}^{xz}(\mathbf{C}^{z})^{-1}\mathbf{C}^{zx} \ .
\end{aligned}$$

The second and third terms are independent of $\mathbf{K}$. As $\mathbf{C} \geq \mathbf{0} \Rightarrow \mathbf{T}\mathbf{C}(\mathbf{T})^\top \geq \mathbf{0}$ for arbitrary matrices $\mathbf{T}$, the first term is positive semi-definite. Therefore, it is minimized when $\mathbf{K} = \mathbf{C}^{xz}(\mathbf{C}^{z})^{-1}$. □

Interestingly, there is a close relation between the orthogonality of measurements and LMMSE estimators. In fact, an equivalence between these two concepts can be proven.

Theorem 2.4 (Weak Orthogonality Principle) *Let $\underline{\mathbf{z}}$ denote a random measurement vector. The estimator $\underline{\hat{\mathbf{x}}}$ is the LMMSE estimator (2.17), if and only if $\mathrm{E}\{\underline{\mathbf{x}}\} = \mathrm{E}\{\underline{\hat{\mathbf{x}}}\}$ and $\mathrm{E}\{\underline{\mathbf{z}}(\underline{\hat{\mathbf{x}}} - \underline{\mathbf{x}})^\top\} = \mathbf{0}$.*

PROOF. In a first step, let $\hat{\underline{\mathbf{x}}}$ denote the LMMSE estimator. Then, it has already been shown in Lemma 2.2 that the estimator is unbiased. However, for unbiased $\hat{\underline{\mathbf{x}}}$ it holds

$$\mathrm{E}\{\mathrm{E}\{\underline{\mathbf{z}}\}(\hat{\underline{\mathbf{x}}} - \underline{\mathbf{x}})^\top\} = \mathbf{0} \ ,$$

and $\mathrm{E}\{\underline{\mathbf{z}}(\hat{\underline{\mathbf{x}}} - \underline{\mathbf{x}})^\top\} = \mathrm{E}\{(\underline{\mathbf{z}} - \mathrm{E}\{\underline{\mathbf{z}}\})(\hat{\underline{\mathbf{x}}} - \underline{\mathbf{x}})^\top\}$. For the LMMSE estimator (2.17), orthogonality follows with

$$\begin{aligned}\mathrm{E}\{\underline{\mathbf{z}}(\hat{\underline{\mathbf{x}}} - \underline{\mathbf{x}})^\top\} =& \mathrm{E}\{(\underline{\mathbf{z}} - \mathrm{E}\{\underline{\mathbf{z}}\})\cdot \\ & (\mathrm{E}\{\underline{\mathbf{x}}\} - \underline{\mathbf{x}} + \mathbf{C}^{xz}(\mathbf{C}^z)^{-1}(\underline{\mathbf{z}} - \mathrm{E}\{\underline{\mathbf{z}}\}))^\top\} \\ =& \mathbf{C}^{zx} + \mathbf{C}^z(\mathbf{C}^z)^{-1}\mathbf{C}^{zx} = \mathbf{0} \ .\end{aligned}$$

Now, consider an estimator $\hat{\underline{\mathbf{x}}}$ that is unbiased and satisfies the orthogonality property, and let $\tilde{\underline{\mathbf{x}}}$ denote the LMMSE estimator. Then,

$$\begin{aligned}\mathrm{E}\{(\tilde{\underline{\mathbf{x}}} - \underline{\mathbf{x}})^2\} =& \mathrm{E}\{(\tilde{\underline{\mathbf{x}}} - \hat{\underline{\mathbf{x}}} + \hat{\underline{\mathbf{x}}} - \underline{\mathbf{x}})^2\} \\ =& \mathrm{E}\{(\tilde{\underline{\mathbf{x}}} - \hat{\underline{\mathbf{x}}})^2\} + \mathrm{E}\{(\hat{\underline{\mathbf{x}}} - \underline{\mathbf{x}})^2\} + \\ & \mathrm{E}\{(\tilde{\underline{\mathbf{x}}} - \hat{\underline{\mathbf{x}}})(\hat{\underline{\mathbf{x}}} - \underline{\mathbf{x}})^\top\} + \mathrm{E}\{(\hat{\underline{\mathbf{x}}} - \underline{\mathbf{x}})(\tilde{\underline{\mathbf{x}}} - \hat{\underline{\mathbf{x}}})^\top\} \ .\end{aligned}$$

As the difference between two linear estimators is a linear combination, it holds $\tilde{\underline{\mathbf{x}}} - \hat{\underline{\mathbf{x}}} = \tilde{\mathbf{K}}\underline{\mathbf{z}} + \tilde{\underline{\mathbf{m}}}$ for some variables $\tilde{\mathbf{K}}$ and $\tilde{\underline{\mathbf{m}}}$. As the LMMSE estimator is unbiased according to Lemma 2.2 and $\hat{\underline{\mathbf{x}}}$ is unbiased according to the assumptions, $\tilde{\underline{\mathbf{m}}}$ is a linear transformation of the expected measurement vector, which is uncorrelated from $\underline{\mathbf{x}}$. With the orthogonality of $\hat{\underline{\mathbf{x}}}$, it follows

$$\mathrm{E}\{(\tilde{\mathbf{K}}\underline{\mathbf{z}} + \tilde{\underline{\mathbf{m}}})(\hat{\underline{\mathbf{x}}} - \underline{\mathbf{x}})^\top\} = \tilde{\mathbf{K}}\,\mathrm{E}\{\underline{\mathbf{z}}(\hat{\underline{\mathbf{x}}} - \underline{\mathbf{x}})^\top\} + \mathrm{E}\{\tilde{\underline{\mathbf{m}}}(\hat{\underline{\mathbf{x}}} - \underline{\mathbf{x}})^\top\} = 0 \ .$$

Therefore, it holds $\mathrm{E}\{(\tilde{\underline{\mathbf{x}}} - \underline{\mathbf{x}})^2\} = \mathrm{E}\{(\tilde{\underline{\mathbf{x}}} - \hat{\underline{\mathbf{x}}})^2\} + \mathrm{E}\{(\hat{\underline{\mathbf{x}}} - \underline{\mathbf{x}})^2\}$. With $\mathrm{E}\{(\tilde{\underline{\mathbf{x}}} - \hat{\underline{\mathbf{x}}})^2\} > \mathbf{0}$ for $\hat{\underline{\mathbf{x}}} \neq \tilde{\underline{\mathbf{x}}}$, it follows $\tilde{\underline{\mathbf{x}}} = \hat{\underline{\mathbf{x}}}$. □

For a discussion on the processing of several measurements, the reader is referred to [128]. Indeed, as several measurements can be lumped together to form an augmented measurement, Lemma 2.3 and Theorem 2.4 apply naturally. The inverse problem, i.e., one or several measurements belong to more than one state, is examined under the topic track association [34] and constitutes a key component of multi-target tracking [160].

Finally, it is worth pointing out the relation between the different estimators. As shown in Theorem 2.1, the MMSE estimator is the conditional mean and minimizes the MSE for all measurements. The analogous concept in frequentist statistics is the minimum-variance unbiased estimator (MVUE).

The LMMSE estimator is the best linear estimator of the form $\hat{\underline{\mathbf{x}}} = \underline{\mathbf{m}} + \mathbf{K}\underline{\mathbf{z}}$. As the minimization holds in the positive definite sense, derivations of the LMMSE estimator that utilize criteria such as trace, determinant, or projections of the MSE matrix lead to the same estimator. However, for specific measurements, better estimators can be derived. The counterpart in frequentist statistics is the best linear unbiased estimator (BLUE).

In general, the MSE of the MMSE estimator is smaller than the one of the LMMSE estimator. For **Gaussian densities**, however, it has been shown that LMMSE and MMSE estimator coincide [47, 128]. Therefore, the "optimal" estimator can also be derived by exploiting the representation of Gaussian densities [47].

2.2.2 Dynamic State Estimation

The combination of measurements as it has been discussed so far is referred to as static estimation. Now, states that evolve according to the state models from Section 2.1 are considered. For this purpose, let the system and measurement models satisfy the assumptions from Definition 2.1. The objective is to obtain the LMMSE estimator for dynamic states, i.e., to derive rules for the combination and processing of estimates and measurements that minimize the MSE matrix. If not otherwise stated, dynamic systems are considered for an arbitrary time period. Therefore, throughout the thesis, let $\mathcal{K} \subseteq \mathbb{N}_0$ denote a potentially unbounded set of time indices and $k \in \mathcal{K}$ a time step. For the sake of consistency and simplicity, estimators $\hat{\underline{\mathbf{x}}}$ are construed as random vectors with covariance $\mathbf{C} = \mathrm{E}\{(\hat{\underline{\mathbf{x}}} - \underline{\mathbf{x}})^2\}$.

It is a direct consequence of the definition and demonstrated in the following that in the considered linear framework the covariance of an estimator depends only on the transformations and covariances of noise terms. Hence, the challenge of finding the LMMSE estimator boils down to the derivation of transformation rules that minimize $\mathbf{C}$. Then, the processing of estimates and measurements is already uniquely determined.

The considerations are confined to the **recursive** processing of estimates and covariances, i.e., to estimators that use previous outputs as inputs in subsequent time steps. This is a reasonable constraint from a practical perspective as recursive algorithms are associated with limited computational effort, and it is justified theoretically, as the LMMSE estimator for the considered linear system is recursive [93]. More details on the latter aspect are given at the end of this section. For recursive estimators only four operations must be considered that cover the challenges in ordinary estimation problems. In the following, the processing of estimates and covariances in initialization, prediction, filtering, and fusion operations are examined. For the sake of simplicity, covariance formulas are given without considering potential bias terms. Therefore, the formulas are only exact when the estimator is unbiased.

Usually an estimate $\tilde{\underline{\mathbf{x}}}_0$ with covariance $\tilde{\mathbf{C}}_0$ is provided at **initialization**. For specific estimators, this requirement can be relaxed by assuming an infinite uncertainty of the initial estimate, which results in a negligence of initial information. In order to cover such cases, initialization operations of the form

$$\hat{\underline{\mathbf{x}}}_0 = \mathbf{T}\tilde{\underline{\mathbf{x}}}_0 \,, \tag{2.19}$$

$$\mathbf{C}_0 = \mathbf{T}\tilde{\mathbf{C}}_0(\mathbf{T})^\top \ , \tag{2.20}$$

with initial transformation matrix $\mathbf{T} \in \mathbb{R}^{n_x \times n_x}$ are considered. Then, a transformation in the information space [116] can for example be realized with $\mathbf{T} = (\tilde{\mathbf{C}})^{-1}$. With state transition matrix $\mathbf{A}_k$ and process noise covariance $\mathbf{Q}_k$, the state **prediction** from (2.8) is reflected in the estimation process by

$$\hat{\underline{\mathbf{x}}}_{k+1} = \mathbf{A}_k \hat{\underline{\mathbf{x}}}_k \ , \tag{2.21}$$

$$\mathbf{C}_{k+1} = \mathbf{A}_k \mathbf{C}_k (\mathbf{A}_k)^\top + \mathbf{Q}_k \ . \tag{2.22}$$

In fact, other transformations are conceivable as well. However, for transformations with $\tilde{\mathbf{A}}_k \neq \mathbf{A}_k$, the estimator is biased, which is suboptimal according to Lemma 2.2. Note that due to the uncertainty in the evolution of the true state, the covariance of the estimate $\mathbf{C}_k$ is increased by the process noise covariance in each time step. Hence, estimators in dynamic state estimation must incorporate system outputs as otherwise the covariance diverges for general systems. Let $\underline{\mathbf{z}}_k$ denote the noisy system output from (2.9). Then, the **filtering** operation describes a linear combination rule for estimate and measurement of the form

$$\hat{\underline{\mathbf{x}}}_{k|k} = \mathbf{L}_k \hat{\underline{\mathbf{x}}}_k + \mathbf{K}_k \underline{\mathbf{z}}_k \ , \tag{2.23}$$

$$\mathbf{C}_{k|k} = \mathbf{L}_k \mathbf{C}_k (\mathbf{L}_k)^\top + \mathbf{K}_k \mathbf{R}_k (\mathbf{K}_k)^\top \ , \tag{2.24}$$

with matrices $\mathbf{L}_k \in \mathbb{R}^{n_x \times n_x}$ and $\mathbf{K}_k \in \mathbb{R}^{n_x \times n_z}$. Taking into account Lemma 2.2, unbiased combinations are of particular interest. It is easy to verify with basic linear algebra that the combination is unbiased for $\mathbf{L}_k = (\mathbf{I} - \mathbf{K}_k \mathbf{H}_k)$, which leads to the filter equation

$$\hat{\underline{\mathbf{x}}}_{k|k} = \hat{\underline{\mathbf{x}}}_k + \mathbf{K}_k (\underline{\mathbf{z}}_k - \mathbf{H}_k \hat{\underline{\mathbf{x}}}_k) \ . \tag{2.25}$$

Note that the filter result is denoted with the subscript $_{k|k}$ to indicate that information of time step k has been comprised. The general subscript $_k$ (without $_{|k}$) is used for either filtered or unfiltered variables.

In sensor networks, estimates from neighbors $\mathcal{S}^s \subseteq \mathcal{S}$ are received at sensor s. For the sake of a simple notation, no distinction is made between $\mathcal{S}^s$ and $\mathcal{S}$, and fused estimates are denoted without a sensor index. The challenge is to derive **fusion** gains $\mathbf{F}$ for the combination according to

$$\underline{\hat{\mathbf{x}}} = \sum_{s \in \mathcal{S}} \mathbf{F}^s \underline{\hat{\mathbf{x}}}^s \,, \tag{2.26}$$

$$\mathbf{C} = \sum_{s,\tilde{s} \in \mathcal{S}} \mathbf{F}^s \mathbf{C}^{s\tilde{s}} (\mathbf{F}^{\tilde{s}})^\top \,, \tag{2.27}$$

where $\mathbf{F}^s = \mathbf{0}$ for $s \notin \mathcal{S}^s$ and the combination is unbiased for $\sum_{s \in \mathcal{S}} \mathbf{F}^s = \mathbf{I}$. A more detailed discussion on the fusion of estimates is given in Section 2.3. The processing of recursive linear estimators is summarized in Algorithm 2.1.

Algorithm 2.1 Recursive Linear Estimator

Initialization: $\underline{\hat{\mathbf{x}}}_0 = \mathbf{T}\underline{\tilde{\mathbf{x}}}_0$
for $k = 1; k \in \mathcal{K}; k = k + 1$ **do**
    Prediction: $\underline{\hat{\mathbf{x}}}_k = \mathbf{A}_{k-1}\underline{\hat{\mathbf{x}}}_{k-1}$
    Filtering: $\underline{\hat{\mathbf{x}}}_{k|k} = \mathbf{L}_k\underline{\hat{\mathbf{x}}}_k + \mathbf{K}_k\underline{\mathbf{z}}_k$
    Fusion: $\underline{\hat{\mathbf{x}}} = \sum_{s \in \mathcal{S}} \mathbf{F}^s \underline{\hat{\mathbf{x}}}^s$
end for

From (2.24) and (2.27), the need to calculate covariances becomes apparent. Apart from serving as quality measure, for example utilized in control applications to determine the force of control actions, covariances are needed to optimize gains $\mathbf{L}$, $\mathbf{K}$, and $\mathbf{F}$. A formalization of linear estimators for dynamic systems is given next.

Definition 2.3 *A recursive* ***linear estimator*** *is determined by its operations in initialization* (2.19)*, prediction* (2.21)*, filtering* (2.23)*, and fusion* (2.26) *steps, i.e., by the matrices* $\mathbf{L}_k$*,* $\mathbf{K}_k$*, and* $\mathbf{F}^s_k$*,* $s \in \mathcal{S}$*,* $k \in \mathcal{K}$*.*

Definition 2.3 covers various popular linear estimators that become of interest later on. Distributed algorithms such as consensus filters [124–126] or diffusion filters [26, 27, 30, 78] are special versions of Algorithm 2.1 with scalar fusion weights. The federated Kalman filter [24,25] and the (linear) ensemble Kalman filter [1,46,77] define schemes that can be represented by means of linear transformations. Even covariance intersection [85,87–89] yields linear fusion gains and can be handled in the proposed framework although a nonlinear optimization is necessary to calculate the transformation matrices.

2.2.2 - a The Kalman Filter

Another well-known example that satisfies Definition 2.3 is the linear minimum mean squared error estimator for linear systems that has originally been proposed by Kalman and Bucy for continuous-time and discrete-time systems in [93, 94] and defines recursive rules for the filtering and processing of measurements. The so-called **Kalman filter** (KF) has been derived by means of the (weak) Orthogonality Principle (Theorem 2.4) but is also obtained as the solution to several other optimization criteria as discussed in Section 2.2.1.

For a derivation, let an unbiased estimate with exact covariance be given at initialization. The unbiased prediction (2.21) is defined by the state model. As the measurement noise is uncorrelated from the state according to Definition 2.1, it holds

$$\mathbf{C}^z = \mathrm{E}\{(\mathbf{H}\underline{\mathbf{x}} + \underline{\mathbf{v}})^2\} = \mathbf{H}\mathbf{C}_k(\mathbf{H})^\top + \mathbf{R}$$

and

$$\mathbf{C}^{xz} = \mathrm{E}\{\underline{\mathbf{x}}(\mathbf{H}\underline{\mathbf{x}} + \underline{\mathbf{v}})^\top\} = \mathbf{C}_k(\mathbf{H})^\top \, .$$

Therefore, with Lemma 2.3, the LMMSE filter gain is obtained as

$$\mathbf{K} = \mathbf{C}_k(\mathbf{H})^\top(\mathbf{H}\mathbf{C}_k(\mathbf{H})^\top + \mathbf{R})^{-1} \tag{2.28}$$

with unbiased counterpart $\mathbf{L} = (\mathbf{I} - \mathbf{KH})$. For the LMMSE gains, the covariance formula (2.24) can be simplified to

$$\mathbf{C}_{k|k} = \mathbf{C}_k - \mathbf{K}\mathbf{H}_k\mathbf{C}_k \; . \tag{2.29}$$

Let $\mathbf{K}$ and $\mathbf{L}$ denote the KF gains, $\mathbf{C}_{k|k}$ the covariance from (2.29), and $\tilde{\mathbf{C}}_{k|k}$ a covariance obtained with gains $\tilde{\mathbf{K}} \neq \mathbf{K}$ and $\tilde{\mathbf{L}} \neq \mathbf{L}$. It has already been shown in Lemma 2.3 that the LMMSE filtering yields a covariance that is optimal in the positive semi-definite sense, i.e., $\mathbf{C}_{k|k} \leq \tilde{\mathbf{C}}_{k|k}$. Now, consider prediction and filter equations (2.22) and (2.24), which are bilinear transformations and summations of input covariances. From basic linear algebra, e.g., Observation 7.7.2 in [76], it is known that $\mathbf{C} \leq \tilde{\mathbf{C}} \Rightarrow \mathbf{TC}(\mathbf{T})^\top \leq \mathbf{T}\tilde{\mathbf{C}}(\mathbf{T})^\top$ for arbitrary matrices $\mathbf{T}$. Thus, the covariance after prediction and filter operations is minimal when the input covariance is minimal. The optimality over multiple time steps follows inductively. Hence, the KF is the LMMSE estimator for the considered linear system.

A more detailed derivation and discussion of the KF are given in [168]. It is worth pointing out that an algebraically identical formulation[1] in the **information space** has been derived [116] that permits the initialization of uninformative estimates and the simple filtering of several measurements per time step. A presentation of the KF in the information form and a discussion of its properties are given in the context of distributed estimation in Section 4.1.

2.2.2 - b Nonlinear Models

The KF has been derived on the premise that underlying system and measurement models are linear. However, the LMMSE theory can also be applied to nonlinear systems by deriving the necessary covariances or by linearizing the models [51, 168]. The naïve approach

[1] A conversion between the forms is possible by means of the Woodbury matrix inversion lemma [171].

is to linearize emerging nonlinearities by means of a **Taylor series** around the estimated state. Depending on the number of considered matrix terms, this method is called extended Kalman filter (EKF) or iterated EKF [3, 153]. Indeed, the linearization necessitates the calculation of derivatives and quite often leads to inconsistent and poor results [104]. Hence, linear regression Kalman filters (LRKF) [104], which approximate the transformation of densities with **deterministic sample representations**, have gained importance.

Consider an arbitrary nonlinear function $\underline{\mathbf{y}} = \underline{g}(\underline{\mathbf{x}})$ and let for $L \in \mathbb{N}$ $\{\underline{x}_i\}_{i=1,\ldots,L}$ denote a weighted sample representation of the distribution $\underline{\mathbf{x}}$. The concept of LRKFs is to approximate the distribution $\underline{\mathbf{y}}$ with samples $\underline{y}_i = \underline{g}(\underline{x}_i)$ according to

$$\hat{\underline{y}} = \sum_{i=1}^{L} \omega_i \underline{y}_i \ , \tag{2.30}$$

$$\hat{\mathbf{C}}^y = \sum_{i=1}^{L} \omega_i (\underline{y}_i - \hat{\underline{y}})(\underline{y}_i - \hat{\underline{y}})^\top \ . \tag{2.31}$$

This allows, in particular, the application of the LMMSE estimator from Lemma 2.3 with $\underline{\mathbf{y}} = \underline{\mathbf{z}}$, $\underline{g} \equiv \underline{h}$, and cross-covariance matrix

$$\hat{\mathbf{C}}^{xy} = \sum_{i=1}^{L} \omega_i (\underline{x}_i - \hat{\underline{x}})(\underline{y}_i - \hat{\underline{y}})^\top \ , \tag{2.32}$$

where $\hat{\underline{x}} = \sum_{i=1}^{L} \omega_i \underline{x}_i$. Optimizing choice and weighting of the samples are the main subjects of the LRKF literature. Popular examples are the unscented Kalman filter [43, 86, 166], the divided difference filter [123], the central difference filter [142], and the cubature Kalman Filter [6]. More sophisticated approaches that also allow applicants to specify the number of samples L have been derived by means of the Gaussian filter [16, 79] and the randomized unscented Kalman filter [158]. An implementation that is optimized for a distributed processing of the unscented Kalman filter has been proposed in [103].

Indeed, linearization induces an approximation error so that properties, e.g., the optimality of the KF, do not apply to the entire problem setting but only to the processing after the linearization. As a matter of fact, the nonlinearity and non-Gaussianity of the problem determines whether the application of linearization techniques is meaningful. In the remainder of this thesis, only the resulting linear(-ized) systems are considered.

2.3 Estimation in Sensor Networks

With information stemming from spatially distributed sources and the amount of information exceeding computational capabilities of single computing devices, there comes a need for parallel computing and distributed processing architectures. Hence, over the last decades, (wireless ad hoc) sensor networks have gained importance in theoretical and practical works.

The idea is to consider several nodes or sensors that take local measurements and exchange information in a network. Throughout the thesis, let $\mathcal{S} = \{s_1, \dots, s_S\}$ denote the set of interconnected nodes that constitute the sensor network and let $s, \tilde{s} \in \mathcal{S}$ denote arbitrary nodes from this set. The nodes can either be pure sensors, which only observe and communicate measurements, "smart" processing nodes [143, 144], or a combination of both. Depending on the scenario, the objective is to optimize the estimate at one fusion center or the estimates at several nodes.

The main difference between central and sensor network schemes is the estimators' access to information. Data vectors such as measurements and initial estimates are instances of random variables and are only known to the sensor that observes them. For the exchange of local information, a transmission between sensors over the underlying network is necessary. Typically, an abstraction of the network is used that enables the modeling of limited capacity, packet collisions, etc. as stochastic or deterministic **packet losses** and **packet delays**.

In the most general form of sensor network estimation, filter and fusion operations are optimized and carried out at local nodes without information about values of remote nodes. System and remote measurement models, however, might be known by the sensors. In fact, the system model (2.8) describes the evolution of the state. Therefore, it must be known to all processing nodes to predict local estimates.

Definition 2.4 System model knowledge *is available in a sensor network, if the system model* (2.8) *is known to all nodes* $s \in \mathcal{S}$.

Measurement models (2.9) define the relation between state and local measurements. Depending on whether the sensor models are nonlinear or state-dependent, they can be assumed known to remote nodes.

Definition 2.5 Global measurement model knowledge *is defined to be available, if sensors have access to the measurement models* (2.9), *i.e., the matrices* $\mathbf{H}_k^s$ *and* $\mathbf{R}_k^s$ *of all nodes* $s \in \mathcal{S}$ *for all* $k \in \mathcal{K}$.

A more comprehensive discussion about model knowledge is given in Section 2.4.

Other criteria for the classification of sensor networks pertain to the communication of nodes. For example, different **network topologies** are distinguished. If the communication between nodes is unrestricted, it proves difficult to maintain dependency information between estimates. Hence, special types such as linear, hierarchical, or fully connected networks are considered in literature [107,108]. In addition, decompositions of the network into subsystems can facilitate the estimation process [97].

Criteria	Possible Values
Node types	Sensors, local (pre-)processors, fusion center
Network characteristics	Packet delays, packet losses
Communication type	Deterministic, stochastic, unforeseeable
Packet type	Measurements, estimates, bundles
Model knowledge	System model, local/global measurement models
Network topology	Fully connected, linear, tree, decompositions

Table 2.1: Criteria for the classification of sensor networks.

Even though a comprehensive characterization of sensor networks is hardly possible, Table 2.1 gives a notion of multiple parameters of sensor networks that affect the performance of estimation algorithms. A more elaborate discussion on the distinction between different sensor network types is given in [33].

Estimators developed in the context of sensor networks are applicable to a wide range of problems that are not necessarily confined to the composition of small sensor devices. For example, the algorithms are prominent in surveillance and target tracking literature [11, 107], where information from satellites, radars, airplanes, etc. must be incorporated. The theory is also relevant for all types of problems that involve an unreliable network such as car-to-car communication [45]. Moreover, the combination of estimation and control with the objective to cooperatively sense and impair the environment has attracted considerable attention under the term sensor-actuator networks [65], [175].

The remainder of this section is devoted to specific sensor network setups. While a unique nomenclature is not used in literature, a plausible differentiation between the three processing schemes centralized, distributed, and decentralized estimation has been proposed [62, 137, 138] and is depicted in Figure 2.1.

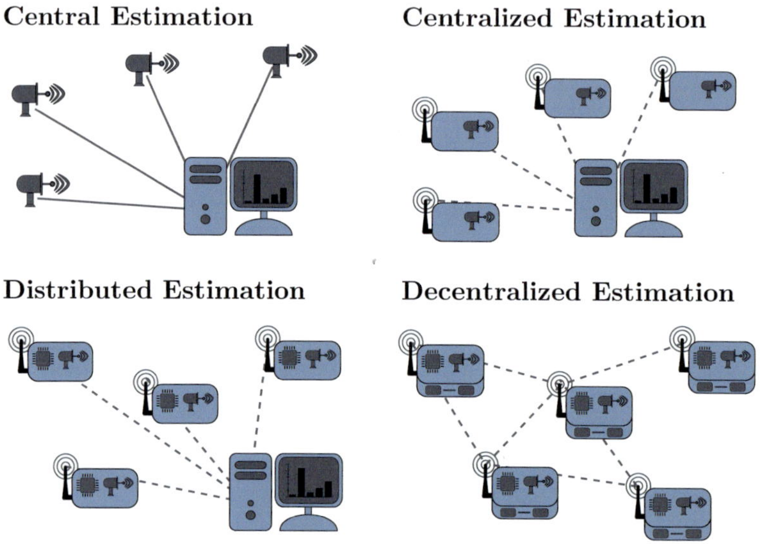

Figure 2.1: Different types of estimation schemes for sensor networks. Solid lines depict reliable and dashed lines unreliable (often wireless) communication links. The central fusion center is illustrated as workstation while sensors can be equipped with communication, computation, and storage devices.

The central scheme exhibits fixed connections between sensor devices and the fusion center. In centralized estimation, measurements are communicated via a network. Smart sensors that transmit preprocessed estimates in a potentially hierarchical network to a fusion center form the distributed estimation scheme. Decentralized estimation constitutes the most general concept where nodes communicate, process, and store data locally.

2.3.1 Centralized Estimation

In centralized estimation, sensors communicate locally observed measurements to a dedicated **fusion center** that filters the data and calculates an estimate. The sensors are assumed to be equipped with no or low computing power and are subject to energy constrains. The fusion center does not have such constraints and is (solely) supposed to provide the sensor network estimate.

For the linear system from Definition 2.1, the centralized LMMSE estimator consists of a buffer in which all measurements with the corresponding models are stored and a central KF [93] that computes an estimate based on all received data. A simple extension is to distribute the calculations of the central KF by transforming measurements locally into the information space before transmitting the data to the fusion center [42]. This technique is meaningful when the measurement space is greater than the state space and reduces the computational effort at the fusion center.

The LMMSE scheme works well when no packet delays or losses occur. In fact, if the communication is not delayed, the buffer is superfluous and the LMMSE estimator can even be calculated recursively. However, otherwise, the theoretic baseline for centralized and distributed estimators in linear systems is only achieved by transmitting the history of all locally observed measurements together with the corresponding models in every communication cycle. Obviously, this scheme requires infinitely growing computational and communication effort.

Analyzing the stochastic properties of the communication allows the derivation of criteria for the **stability** of the centralized estimator subject to the packet loss probability [154]. Moreover, the expected covariance can be bounded. A generalization of the results to partial observation losses, i.e., measurements are only partially lost due to the communication, has been presented in [109]. Further results pertaining to stochastic packet delays have been achieved in [141, 143, 144] by deriving optimal constant gain estimators and providing stability criteria and expected covariance formulas.

Depending on the considered maximum packet delay and the number of sensors, computational effort and required storage of the naïve centralized KF scheme can be considerable. Hence, **out-of-sequence measurement** (OOSM) techniques have been proposed that permit the inclusion of delayed measurements into recursively obtained estimates of the latest time step [8]. Extensions to multi-step lags [10] and augmented measurement vectors [22] have been proposed to improve the results. Optimal OOSM estimators can be achieved, e.g., by means of accumulated state densities [57, 58].

Although the centralized processing scheme captivates with simplicity and theoretical results on stability, the computational effort is bundled at one node and the performance suffers from packet losses, as information is irrevocably lost when a measurement does not arrive at the fusion center. Even if the performance degradation can be cushioned by employing acknowledgement protocols such as TCP [133], such remedies induce additional delays and increase the systems' complexity so that a local preprocessing of measurements as discussed in the following is often advantageous.

2.3.2 Distributed Estimation

In distributed estimation, sensors are equipped with computational power and generate **tracks** from locally observed measurements. The tracks are communicated to a dedicated fusion center and are combined there to obtain an estimate of the state. For the generation of tracks, sensors employ a preprocessing consisting of the filtering and prediction of measurements that allows compressing information in small packets, e.g., in recursively obtained estimates. Then, if the fusion center requests sensor information, only one vector per node that contains the information of all locally observed measurements must be communicated to the fusion center. In particular, packet losses are compensated since subsequent transmissions contain potentially lost information. In order to minimize the estimation error at the fusion center, the processing and fusion of these tracks must be optimized, i.e., the **track-to-track fusion** (T2TF) problem must be solved. Indeed, as already mentioned, the generation of local estimates requires system knowledge from Definition 2.4 to be available at the sensors.

For uncorrelated data, it is sufficient to optimize the processing separately at each node and to combine the estimates with a convex combination to minimize the error [34]. Having said this, in dynamic state estimation, correlations emerge between estimates. This phenomenon, referred to as **common process noise** [7, 9], describes dependencies between estimates due to a common uncertainty in the evolution of the state that affects all estimates.

In order to illustrate the implications, consider the centralized KF from Section 2.3.1 that is obtained by transmitting measurements via a reliable communication to the fusion center. This estimator is not subject to common process noise and calculates the LMMSE estimate. Now, consider a simple scheme where two sensors process measurements with local KFs and their estimates are fused after several time steps at the fusion center. Although (or actually because) the processing is locally optimal and even if the LMMSE combination [9] is used to combine the estimates at the fusion center, the fused estimate is – in general – different from the one of the (optimal) centralized KF [28]. A numerical illustration of this problem will be provided in Example 4.1.

In fact, LMMSE estimation for sensor networks with local preprocessing involves the concurrent optimization of the filter processing of all nodes. Various techniques such as equivalent measurements [34] or the tracklet processing [41] have been proposed that permit the generation of local tracks and yield only slightly suboptimal results at the fusion center.

Still, filter gains in the aforementioned approaches are optimized according to local covariances. A fundamentally different approach is to optimize the processing at sensor nodes premised on the assumption that the estimates are combined at the fusion center subsequently. This concept has been established with the federated Kalman filter [24,25] that achieved **globally optimized filter gains** by means of artificially inflated covariances.

More precisely, covariances of local estimates are obtained from definition but with process noise covariances that are inflated by scalar factors. As a result, covariances and filter gains at the sensors resemble the ones of a centralized KF, and the covariance of the estimate at the fusion center can be easily bounded. An extension of this idea to nonlinear systems has been proposed in [176].

If distributed filters are exclusively optimized according to the fusion center, **decompositions of the centralized KF** are obtained. While parallelized architectures that provide LMMSE estimates when a communication between sensors and fusion center is established in every time step have been proposed in [73,137,170], a formulation that allows combining measurements from several time steps and that is algebraically identical to the centralized KF has been derived in [100]. Generalizations of the exact T2TF [100] to the distributed Kalman filter that permits the inclusion of estimates from arbitrary many sensors and provides simple formulas for arbitrary densities have been proposed in [56, 59, 60].

Moreover, distributed estimation with intermittent transmissions has attracted attention. In [75, 172], system and measurement models were assumed time-invariant and packet delays were neglected. Based on these assumptions, stability criteria for Poisson distributed packet losses could be derived. In [63], optimality and stability theorems have been obtained for (specific) simple estimation networks that employ local Kalman filters. In the consensus framework with scalar fusion weights, distributed estimation has been examined in [23].

2.3.3 Decentralized Estimation

The concurrent state estimation at several nodes is referred to as decentralized estimation. It encompasses the local filtering of measurements and the fusion of exchanged estimates. Information is spread through the network in a hop-to-hop communication. In literature, estimators for such scenarios are also termed gossip algorithms [39]. Note that distributed estimation is a special case, namely when the estimate is evaluated at only one node. Therefore, the challenges from distributed estimation, i.e., in particular, the common process noise, directly transfer to decentralized estimation.

When nodes repeatedly exchange estimates or when common information is spread through different communication paths, dependencies arise between the estimates in the network. This phenomenon is referred to as **common prior information** [36] and is illustrated in Figure 2.2. A popular technique to cope with those dependencies is the explicit bookkeeping of data transfers. For specific communication topologies, e.g., for trees, an information graph can be maintained that enables the "decorrelation" of tracks [34, 107]. For an arbitrary communication, the Channel filter [44,112,113] has been derived to facilitate the bookkeeping of past data exchanges. However, the effort of applying the Channel filter is considerable, as variables for one-to-one relations have to be stored between all nodes in the sensor network. Alternative approaches that bypass the explicit reconstruction of dependency information by bounding the covariance of the fused estimate are discussed in Section 3.2.

Deriving the LMMSE estimator for decentralized systems requires an optimization criterion that involves the estimates at several nodes. The natural extension of the LMMSE criterion is the optimization of the average of local MSEs. However, the minimization of the average or sum of local MSEs proves difficult. As a matter of fact, it has been shown that for the corresponding control problem even in a simple setup with two nodes the separation principle does not hold [32] and

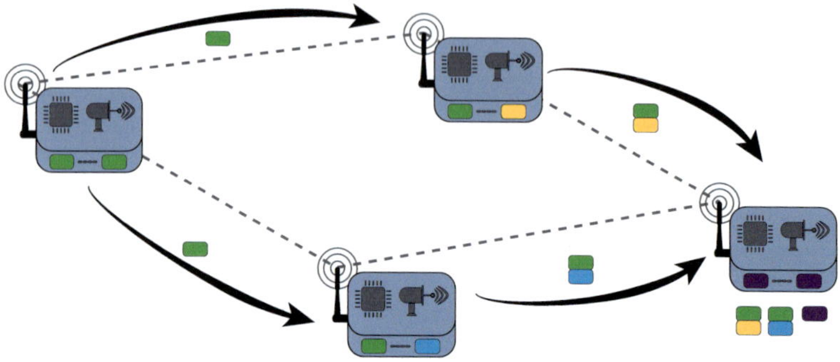

Figure 2.2: An illustration of the emergence of common prior information in decentralized estimation. Due to ambiguous communication paths, the same (green) information is aggregated in several estimates.

the optimization of feedback gains for the general decentralized control problem is NP hard [19]. Therefore, popular approaches focus on static estimation, scalar fusion weights, or certain types of filter and fusion gains.

Decentralized estimation of a deterministic parameter that does not evolve according to a dynamic model provides the basis for the consensus, incremental subgradient, and diffusion approaches discussed below. An overview over recent results on this field with the focus on convergence speed and quantization is given in [39]. For example, the distributed calculation of the centralized maximum likelihood estimator and the best linear unbiased estimator subject to noisy communication links is examined in [145].

Consensus approaches [124–126] synchronize the estimates at the nodes, e.g., by calculating the mean of the initial values. Thus, the focus of consensus protocols is on convergence speed and the derivation of permissible protocols to achieve a consensus rather than the minimization of the MSE. Still, the consensus KF [125] optimizes filter gains by means of the measurement models of the nodes' neighborhood; therefore, it implements ideas of efficient distributed estimation. An examination of the interplay between dynamic state estimation and consensus is provided in [23]. The concurrent optimization of distributed estimation and motion control of mobile sensor devices in the consensus framework has been examined in [127].

A fundamentally different approach is to construct a communication path that contains all nodes. Then, an estimate is communicated along this path and is adjusted at the nodes according to local information. When the overall cost function is the sum of convex local functions, this can be construed as an instance of an **incremental subgradient algorithm** [98,117]. Although the adaptions to the estimate are calculated according to local cost functions, criteria for the convergence speed, i.e., the number of necessary cycles depending on the step size can be given. The optimization of quantized estimates in this framework has been studied in [136]. Indeed, a unique path through the network is not only difficult to find in sensor networks but is also prone to node failures or packet losses.

Therefore, the concurrent communication of all nodes is considered in **diffusion approaches** [26,27,30,78]. The idea is to fix the filter processing and to optimize the convex combination of estimates by means of a diffusion matrix. Properties for scalar fusion weights, in particular, concerning the steady state performance are derived in [26]. An extension of the results to smoothing is given in [27]. Other works are concerned with the learning of diffusion matrices [30] or the guarantee of stable estimates [78]. As, in the considered diffusion framework, estimates are not imposed to converge to a consensus, the MSE of diffusion approaches is in general smaller than the one of consensus approaches [26].

Still, research in decentralized estimation is far away from being regarded as complete. In particular, the presented algorithms do not even provide the LMMSE estimates for dynamic state estimation when they are applied to the special case of distributed estimation.

2.3.4 Joint Space Covariance Representation

The challenges of LMMSE estimation in distributed and decentralized estimation are illustrated by an augmentation of sensor estimates. The joint space is a comprehensive representation for the processing of multiple estimates in sensor networks. For a depiction consider an omniscient observer that has access to the measurements, estimates, and processing matrices of all sensors. Then, an augmentation of all estimates in the sensor network into one joint vector permits the unified representation of local covariances and cross-covariance matrices in form of the **joint covariance matrix**

$$\bar{\mathbf{C}}_k = \begin{pmatrix} \mathbf{C}_k^{s_1} & \dots & \mathbf{C}_k^{s_1 s_S} \\ \vdots & \ddots & \vdots \\ \mathbf{C}_k^{s_S s_1} & \dots & \mathbf{C}_k^{s_S} \end{pmatrix} = \mathrm{E}\left\{ \left(\begin{pmatrix} \hat{\underline{\mathbf{x}}}_k^{s_1} \\ \vdots \\ \hat{\underline{\mathbf{x}}}_k^{s_S} \end{pmatrix} - \begin{pmatrix} \underline{\mathbf{x}}_k \\ \vdots \\ \underline{\mathbf{x}}_k \end{pmatrix} \right)^2 \right\} . \quad (2.33)$$

The local operations of linear estimators, which are specified in Algorithm 2.1, are given by block diagonal transformations in the joint space. As the process noise reflects an uncertainty in the evolution of the state that is common to all estimates, the predicted joint covariance matrix is obtained with

$$\bar{\mathbf{C}}_{k+1} = \begin{pmatrix} \mathbf{A}_k & \dots & \mathbf{0} \\ \vdots & \ddots & \vdots \\ \mathbf{0} & \dots & \mathbf{A}_k \end{pmatrix} \bar{\mathbf{C}}_k \begin{pmatrix} (\mathbf{A}_k)^\top & \dots & \mathbf{0} \\ \vdots & \ddots & \vdots \\ \mathbf{0} & \dots & (\mathbf{A}_k)^\top \end{pmatrix} + \begin{pmatrix} \mathbf{Q}_k & \dots & \mathbf{Q}_k \\ \vdots & \ddots & \vdots \\ \mathbf{Q}_k & \dots & \mathbf{Q}_k \end{pmatrix} . \quad (2.34)$$

For independent measurement noise terms and with local filter gains $\mathbf{L}_k^s$ and $\mathbf{K}_k^s$, the transformation matrices as well as the joint measurement noise matrix are block-diagonal. Hence, the joint covariance matrix of the filtered estimates is obtained according to

$$\bar{\mathbf{C}}_{k|k} = \begin{pmatrix} \mathbf{L}_k^{s_1} & \dots & \mathbf{0} \\ \vdots & \ddots & \vdots \\ \mathbf{0} & \dots & \mathbf{L}_k^{s_S} \end{pmatrix} \bar{\mathbf{C}}_k \begin{pmatrix} (\mathbf{L}_k^{s_1})^\top & \dots & \mathbf{0} \\ \vdots & \ddots & \vdots \\ \mathbf{0} & \dots & (\mathbf{L}_k^{s_S})^\top \end{pmatrix} + \\ \begin{pmatrix} \mathbf{K}_k^{s_1} & \dots & \mathbf{0} \\ \vdots & \ddots & \vdots \\ \mathbf{0} & \dots & \mathbf{K}_k^{s_S} \end{pmatrix} \begin{pmatrix} \mathbf{C}_k^{s_1} & \dots & \mathbf{0} \\ \vdots & \ddots & \vdots \\ \mathbf{0} & \dots & \mathbf{C}_k^{s_S} \end{pmatrix} \begin{pmatrix} (\mathbf{K}_k^{s_1})^\top & \dots & \mathbf{0} \\ \vdots & \ddots & \vdots \\ \mathbf{0} & \dots & (\mathbf{K}_k^{s_S})^\top \end{pmatrix} . \tag{2.35}$$

As it becomes apparent from (2.35), filter operations are block diagonal transformations of joint covariance matrices. However, the multiplication of the joint covariance matrix with block diagonal transformation matrices does not involve the summation of matrix blocks $\mathbf{C}_k^{s\tilde{s}}$, as it would be the case for fully occupied transformation matrices. In fact, matrices $\mathbf{K}_k^s$ and $\mathbf{L}_k^s$ transform only the s-th (block) row and column. Therefore, when the sum of the MSEs of the estimates, i.e., the trace of $\bar{\mathbf{C}}$, is taken as the optimization criterion, the optimal gains are the ones that minimize the local covariances $\mathbf{C}_k^s$. Indeed, as discussed in Section 2.2.2 - a, these gains are obtained by means of **local KFs**.

The exchange and linear fusion of estimates can be represented in the joint space as well. To this end, let $\mathbf{F}_{\tilde{s}}^s$ denote the fusion gain by which the estimate from sensor s is fused into the estimate at sensor $\tilde{s}$, i.e., $\hat{\underline{\mathbf{x}}}^{\tilde{s}} = \sum_{s \in \mathcal{S}} \mathbf{F}_{\tilde{s}}^s \hat{\underline{\mathbf{x}}}^s$. Then, the fused joint covariance matrix is obtained according to

$$\bar{\mathbf{C}} = \begin{pmatrix} \mathbf{F}_{s_1}^{s_1} & \dots & \mathbf{F}_{s_1}^{s_S} \\ \vdots & \ddots & \vdots \\ \mathbf{F}_{s_S}^{s_1} & \dots & \mathbf{F}_{s_S}^{s_S} \end{pmatrix} \bar{\mathbf{C}} \begin{pmatrix} (\mathbf{F}_{s_1}^{s_1})^\top & \dots & (\mathbf{F}_{s_S}^{s_1})^\top \\ \vdots & \ddots & \vdots \\ (\mathbf{F}_{s_1}^{s_S})^\top & \dots & (\mathbf{F}_{s_S}^{s_S})^\top \end{pmatrix} , \tag{2.36}$$

where $\mathbf{F}_{\tilde{s}}^s = \mathbf{0}$, if the estimate s is not received by node $\tilde{s}$ and $\sum_{s \in \mathcal{S}} \mathbf{F}_{\tilde{s}}^s = \mathbf{I}$ for unbiased combinations of estimates. Again, fusion gains $\mathbf{F}_{\tilde{s}}^s$ of node $\tilde{s}$ only transform the $\tilde{s}$-th row and column of the joint covariance matrix. However, in contrast to prediction and filter operations, the cross-covariance matrices $\mathbf{C}_k^{s\tilde{s}}$ from the off-diagonal blocks affect the fused covariances. For a given joint covariance matrix $\bar{\mathbf{C}}$, the **LMMSE fusion gains** at node $\tilde{s}$ have been derived for two [9] and arbitrary many [2, 148, 161, 173] estimates.

Theorem 2.5 *Consider estimates* $\hat{\underline{\mathbf{x}}}^s$, $s \in \mathcal{S}$ *with joint covariance matrix* $\bar{\mathbf{C}}$ *from* (2.33). *The optimal linear fusion gains for unbiased estimates* $\hat{\underline{\mathbf{x}}}^s$ *are given by*

$$\mathbf{F} = \begin{pmatrix} \mathbf{F}^{s_1} & \dots & \mathbf{F}^{s_S} \end{pmatrix} = \mathbf{C}(\underline{\mathbf{1}})^\top (\bar{\mathbf{C}})^+ , \tag{2.37}$$

where $\mathbf{C} = \left((\underline{\mathbf{1}})^\top (\bar{\mathbf{C}})^+ \underline{\mathbf{1}}\right)^+$ *denotes the covariance of the fused estimate* $\hat{\underline{\mathbf{x}}} = \sum_{s\in\mathcal{S}} \mathbf{F}^s \hat{\underline{\mathbf{x}}}^s$ *and* $\underline{\mathbf{1}} := \begin{pmatrix} \mathbf{I} & \cdots & \mathbf{I} \end{pmatrix}^\top$.

PROOF. The derivation for two estimates is provided in [9] and for an arbitrary number of estimates in [161, 173]. A simple proof that constitutes the result for two estimates in the positive definite sense is provided in [179]. □

Applied to the joint space representation (2.36), the matrix $\mathbf{F}$ from Theorem 2.5 denotes a block row of the transformation matrix. Therefore, the fused covariance from Theorem 2.5 is a short form for the fused block diagonal covariances that are obtained with fusion gains (2.37) by means of (2.36). A special case of Theorem 2.5, namely if the cross-covariance matrices are zero, is the convex combination of estimates.

Corollary 2.6 *Let the estimates* $\hat{\underline{\mathbf{x}}}^s$, $s \in \mathcal{S}$ *be uncorrelated. Then, the optimal fusion gains are given by* $\mathbf{F}^s = \mathbf{C}(\mathbf{C}^s)^{-1}$, *where*

$$\mathbf{C} = \left(\sum_{s\in\mathcal{S}} (\mathbf{C}^s)^{-1}\right)^{-1}$$

denotes the covariance of the fused estimate $\hat{\underline{\mathbf{x}}} = \sum_{s\in\mathcal{S}} \mathbf{F}^s \hat{\underline{\mathbf{x}}}^s$.

The fusion concept can also be generalized to partially overlapping states, i.e., when individual estimates capture only a subset of the entire state space. The corresponding theory has been discussed in the context of control [80], and LMMSE fusion formulas have been given in [122]. Note that Theorem 2.5 and Corollary 2.6 are general fusion techniques that can also be used at the fusion center to combine preprocessed measurements from the sensors.

In the perspective of decentralized estimation, the LMMSE filter operation is given by local KFs and the LMMSE fusion by Theorem 2.5. However, these operations optimize only the average MSE of the local estimators and do not provide a joint covariance matrix that is optimal in the positive definite sense.

Consider for example the filter operation. Even if the filter matrices minimize the individual covariances on the block diagonal in the positive definite sense, no statements can be made about the cross-covariance matrices, which are transformed as well. In particular, it is possible that suboptimal filter gains yield a joint covariance matrix that features an indefinite difference to the joint covariance matrix obtained with local KFs. Then, a subsequent LMMSE fusion can yield smaller covariances on the block diagonal than the ones obtained with the LMMSE filter processing.

The bottom line is that without the positive definite relation, the optimality of the joint covariance matrix after several filter and fusion operations cannot be guaranteed. Therefore, the combination of local KFs with Theorem 2.5 is only **one-step optimal**.

When the models are time-invariant, the multi-step or steady state LMMSE estimators can be obtained as the solutions to optimization problems in the joint space. However, to the author's knowledge, no efficient methods are available to solve the emerging system of nonlinear matrix equations. This applies in particular to sensor networks with limited local knowledge.

2.4 Challenges of LMMSE Estimation in Sensor Networks

The ideas and approaches discussed in the remainder of this thesis touch various research areas and implement different concepts. Still, they all serve to provide LMMSE estimators for sensor networks of linear systems according to Definition 2.1. Of course, this is only the framework in which the estimators are derived and optimized. By means of linearization techniques or simplifications concerning the independence of noise terms, the estimators are applicable to nonlinear systems.

For linear systems, the problem of central estimation can be considered as being solved. The KF constitutes the **LMMSE estimator** and provides covariances that are optimal in the positive definite sense. For distributed and decentralized estimation, efficient estimators have been derived as well. The distributed Kalman filter [56, 59, 60, 100] yields LMMSE estimates and the – at least for one time step – optimal solution to decentralized estimation is obtained by means of local KFs in combination with the LMMSE fusion from Theorem 2.5. However, the distributed Kalman filter requires **global model knowledge**, and in order to apply the LMMSE fusion in decentralized estimation, cross-covariance matrices are necessary that even depend on filter and fusion operations applied at remote nodes.

Indeed, global model knowledge as it is required by both approaches is a rarity in real world systems. Consider for example a distance sensor. It can be expected that uncertainties of measurements increase relative to distances between sensors and object. Thus, an intuitive policy for the sensor is to approximate the noise covariance by means of the estimated distance, which, however, is based on the local estimate of the object's position. Therefore, the measurement model of the distance sensor depends on its local estimate and consequently, is unknown to other nodes in the sensor networks. In fact, the explicit **dependency of model parameters on local estimates** is an innate property of nonlinear models.

Moreover, when the communication is stochastic due to packet delays and losses, only the receiver knows which estimates are processed. In particular, an unbiased combination is only feasible by adapting fusion gains to the number of estimates. Therefore, again, gains depend on realizations of stochastic variables, i.e., on knowledge that is only locally available. Consequently, the **evolution of cross-covariance matrices** cannot be calculated by remote sensors even if the utilized estimation algorithm is known.

Additional constraints are imposed by **limited resources** at the sensors. Especially for large sensor networks or when models are time varying, the effort to store the measurement models of all sensors and the computational costs to process all cross-covariance matrices are too high. However, as discussed in Chapter 1, the challenges of modern sensor networks lie in limited knowledge about sensor network capacities and the need to adapt online to new realities.

The objective of this thesis is to propose techniques for LMMSE estimation in distributed and decentralized estimation that are applicable to modern sensor networks. Hence, the existing approaches for the LMMSE filtering and fusion must be generalized to cope with the aforementioned challenges. In the course of this, two main problems are to be solved. First, efficient techniques for the reconstruction of dependencies between estimates based on locally processed variables must be developed. Then, the LMMSE fusion formula or an adaption of it can be applied. Second, local estimators for the sensors must be derived that operate without global model knowledge and optimize the estimates with respect to the sensor network. Both of the problems are examined in the following and different solutions with various advantages and drawbacks are proposed.

CHAPTER 3

Information Fusion in Sensor Networks

In interconnected estimation systems, information is gathered at spatially distributed processing units. To exploit the potential of these systems, information is exchanged over potentially unreliable communication links. It is a known fact that concurrent communication, especially in wireless networks, poses challenges that manifest themselves in packet delays and losses whose frequency rises with the utilization of network resources. The energy consumption caused by data transmissions thereby largely determines the lifespan of sensors so that in order to minimize data traffic, information should be fused at the earliest possible stage.

Considering the classification of sensor networks into distributed and decentralized estimation schemes, information about a common state from several sensors is either to be weighted at the fusion center or to be integrated regularly into local estimates. Irrespectively of the processing type, the question arises how much trust is to be put in individual estimates in the fusion. A meaningful concept to address this problem in a stochastic setting must involve quality of and dependency between the estimates.

In the framework of linear estimation, covariances and cross-covariance matrices provide a complete specification of the quality of estimates. In fact, the optimal linear combination of estimates is uniquely determined when exact knowledge about the joint covariance matrix is available to the fusing node. However, the quantification of dependencies in sensor networks proves difficult. For an explanation, consider two spatially distributed sensors that are initialized with the same estimate. Let sensor one observe and integrate an independent measurement into its local estimate. Then, the cross-covariance matrix between the sensor estimates depends on the amount of information in the filtered estimate at node one that traces back to the initial value, i.e., on the weighting in the filter step.

An approach that enables the calculation of cross-covariance matrices is to exchange all local transformations along with the estimates and to reconstruct the evolution of covariances at the receiver [99]. Indeed, for multi-hop or irregular communication, this approach is hardly realizable since it requires transmitting huge quantities of data and considerable computational power at the sensors.

As the calculation of cross-covariance matrices poses a major challenge in sensor networks, it is tempting to ignore them in the fusion operation. However, independently of the potentially bad outcome of heuristic fusion methods, the uncertainty of the fused estimate depends on cross-covariance matrices between prior estimates. For a motivation and illustration of the challenges involved in the fusion of two estimates subject to an unknown cross-covariance matrix, consider the next example.

Example 3.1: Motivation of Cross-covariance Matrix Reconstruction

Let estimates $\hat{\underline{\mathbf{x}}}^{s_1}$ and $\hat{\underline{\mathbf{x}}}^{s_2}$ with local covariances $\mathbf{C}^{s_1} = \mathrm{diag}(2,1)$, $\mathbf{C}^{s_2} = \mathrm{diag}(1,2)$ be given and let the cross-covariance matrix $\mathbf{C}^{s_1 s_2}$ be unknown. Then, a linear estimator must determine fusion gains $\mathbf{F}^{s_1}$, $\mathbf{F}^{s_2}$, and the covariance of the fused estimate $\hat{\underline{\mathbf{x}}} = \mathbf{F}^{s_1}\hat{\underline{\mathbf{x}}}^{s_1} + \mathbf{F}^{s_2}\hat{\underline{\mathbf{x}}}^{s_2}$ without knowledge of $\mathbf{C}^{s_1 s_2}$. A naïve approach is to calculate the convex combination of the two estimates [34] with

$$\mathbf{F}^{s} = \left((\mathbf{C}^{s_1})^{-1} + (\mathbf{C}^{s_2})^{-1}\right)^{-1}(\mathbf{C}^{s})^{-1}\ , \quad s \in \{s_1, s_2\}\ .$$

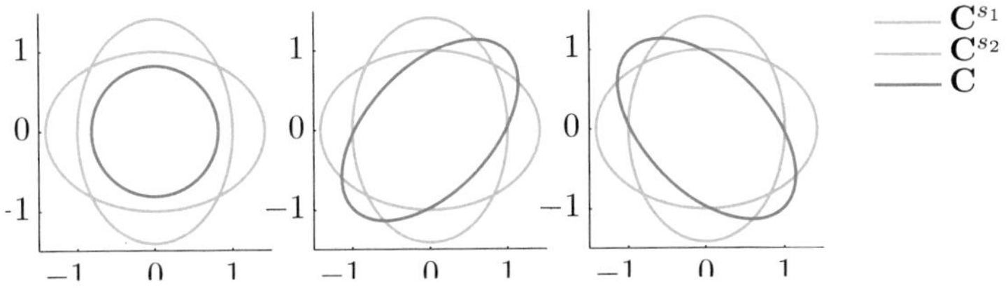

Figure 3.1: Ellipsoids for covariances $\mathbf{C}^{s_1}$ and $\mathbf{C}^{s_2}$ as well as for covariances of fused estimates $\mathbf{C}$ subject to different cross-covariance matrices.

The resulting true covariances of the fused estimate for three possible cross-covariance matrices are depicted in Figure 3.1 by means of the respective covariance ellipsoids $\mathcal{E}(\mathbf{C}) = \{\underline{x}|(\underline{x})^\top(\mathbf{C})^{-1}\underline{x} = 1\}$[1]. Two observations emerge from the data. First, the covariance of the fused estimate depends on the unknown cross-covariance matrix and second, it cannot be stated as a rule that covariances of fused estimates are smaller than prior covariances. ■

It results from Example 3.1 that, even if covariances of estimates prior to the fusion are exactly known, the fusion of estimates subject to limited information about cross-covariance matrices induces uncertainty in the quality assessment of the fused estimate.

Indeed, in decentralized estimation, fused estimates are priors of subsequent fusion operations. Hence, not even local uncertainties could be utilized for the optimization of fusion gains. As covariances of estimates are also required to calculate filter gains and for a potential subsequent use in control or decision problems, the entire information processing becomes questionable when dependency information is neglected.

[1]It holds $\mathbf{C} \leq \mathbf{P}$, if and only if $\mathcal{E}(\mathbf{C}) \subseteq \mathcal{E}(\mathbf{P})$.

Thus, meaningful estimation in modern sensor networks subject to limited information at the sensors is only feasible with techniques that either provide (imprecise) cross-covariance matrices or bypass the dependence on cross-covariance matrices in the fusion operation. In this chapter, both approaches are explored.

In Section 3.1, sample sets are generated from the joint distribution of the error processes and the **joint covariance matrix is estimated** by means of sample covariances. In an analysis of the covariance structure of estimates in linear estimation, it is revealed that generation and processing of samples can be decomposed into local operations so that a scheme to estimate cross-covariance matrices that works based on locally known variables is obtained. Efficient covariance estimators are discussed and the optimal distribution for sampling noise terms in the proposed scheme is derived.

A fundamentally different approach is pursued in Section 3.2 by considering all possible cross-covariance matrices in order to give a worst-case assessment for the covariance of the fused estimate. Hence, the objective is to provide guaranteed **covariance bounds** instead of estimates of covariances that are subject to uncertainty. While covariance bounding under unknown correlations is an established concept, the rigorous generalization to partially known cross-covariance matrices is new and bridges the gap between fusion methods under known and unknown correlations.

It is an insight of the results but also follows intuitively that tightness and precision of the bound on the fused estimate improves relative to additional knowledge about dependencies. Therefore, approaches to reconstruct cross-covariance matrices partially based on different decompositions are derived and evaluated. Finally, an algorithm is proposed that uses a noise decomposition of covariances to achieve arbitrary precision subject to computational and communication effort.

3.1 Sample-based Covariance Estimation

As outlined above, the exact reconstruction of (cross-)covariance matrices in sensor networks is impracticable. In fact, for the naïve technique from [99], the amount of stored and processed data increases with each emerging noise term. However, limiting computational and communication effort is not only a nice feature but also quite often a necessity. In this section, a scheme for the sample-based estimation of covariances that takes into account the limitations of sensor networks is proposed.

The quality of estimators is specified by the distribution of the random **(estimation) error** $\underline{\mathbf{e}} = \hat{\underline{\mathbf{x}}} - \underline{\mathbf{x}}$. Now, consider a state estimator and assume that the distribution of the error is known, e.g., to be a Gaussian density specified by its covariance. Then, samples can be generated from the distribution, which, in turn, form an equally weighted Dirac mixture density and describe the estimation error. In particular, it is possible to maintain the Dirac mixture description of the error density by reflecting the transformations of the estimation error in the sample space. From the obtained representation, properties of the original error density can be reconstructed by means of proper techniques, as it is illustrated in Figure 3.2.

For the considered linear models, the error is linearly combined with independent random noise variables in prediction and filter operations, which corresponds to a convolution of densities. For example, when noise terms are Gaussian distributed, the convolution with Dirac mixtures yields a Gaussian mixture that needs to be approximated by a Dirac mixture to enable further sample-based processing. At this stage, an approximation is necessary. Generally, it is advisable to calculate deterministic sample approximations, as this is more efficient than stochastic sampling [53, 54]. However, deriving Dirac mixtures in sensor networks based on local knowledge without access to the joint error distribution proves difficult.

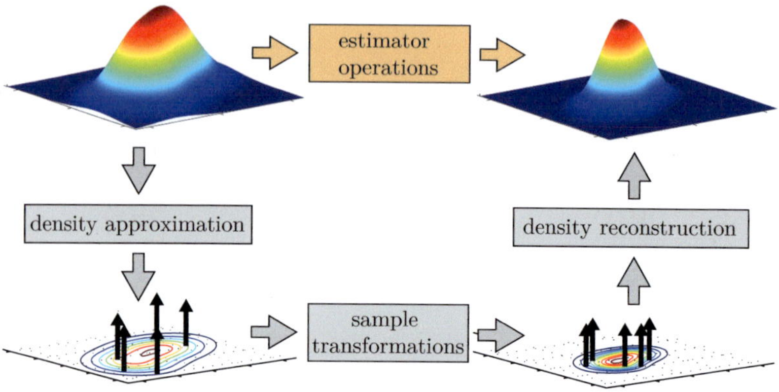

Figure 3.2: A schematic overview describing the transformation of the estimation error and necessary operations to reflect the evolution of the error by means of samples.

Indeed, for estimating cross-covariance matrices in the considered sensor networks, transformations of the underlying state estimator cannot only be assumed linear but also independent from the samples. Therefore, a simple technique to obtain a sample from the error density at an arbitrary time step is to independently generate samples from all occurring noise distributions and to combine these samples according to the noise composition of the estimation error. The generation can be repeated until the desired precision of the density representation is achieved. From the obtained samples, the moments of the error and, in particular, the covariance can be estimated.

In the following, two properties of this simple technique are exploited. First, the proposed sample generation can be calculated recursively such that the sample processing corresponds to a simulation of the estimation error. Second, the generation of joint samples can be distributed to sensors even if the transformations of remote nodes are unknown. For an illustration of the main insights and ideas, the filter operation in a simple network is examined in the next example.

Example 3.2: Introduction to Sample-based Covariance Estimation

Consider a one-dimensional system that is observed by two sensors. The distribution of the joint space error from Section 2.3.4, i.e., of the augmented errors, is two-dimensional and can be approximated by a sample set as depicted in Figure 3.2. When the samples are indexed, two lists with the first, respectively second, coordinate of the samples can be maintained, and by means of the indices, coordinates can be augmented to reconstruct the joint space samples, as it is depicted in the top part of Figure 3.3.

Now, let local estimation errors $\underline{\mathbf{e}}_k^s$ be described by sample sets. When a measurement is filtered, the corresponding error for unbiased estimators with $\mathbf{L}^s = \mathbf{I} - \mathbf{K}^s\mathbf{H}^s$ is given by

$$\begin{aligned}\underline{\mathbf{e}}_{k|k}^s &= \hat{\underline{\mathbf{x}}}_{k|k}^s - \underline{\mathbf{x}}_k \\ &= \mathbf{L}^s\hat{\underline{\mathbf{x}}}_k^s + \mathbf{K}^s\underline{\mathbf{z}}^s - \underline{\mathbf{x}}_k \\ &= \mathbf{L}^s\hat{\underline{\mathbf{x}}}_k^s - (\mathbf{I} - \mathbf{K}^s\mathbf{H}^s)\underline{\mathbf{x}}_k + \mathbf{K}^s\underline{\mathbf{v}}^s \\ &= \mathbf{L}^s\underline{\mathbf{e}}_k^s + \mathbf{K}^s\underline{\mathbf{v}}^s \ .\end{aligned}$$

As discussed above, a simple, recursive processing scheme to obtain a sample representation of the error $\underline{\mathbf{e}}_{k|k}^s$ is to transform the samples of $\underline{\mathbf{e}}_k^s$ with $\mathbf{L}^s$ and to add independently generated noise from the random variable $\mathbf{K}^s\underline{\mathbf{v}}^s$ to each of the samples.

In the next step, let the sample sets be the lists with scalar coordinates of the joint space samples. The concurrent filtering of sensors s_1 and s_2 yields

$$\begin{pmatrix}\underline{\mathbf{e}}_{k|k}^{s_1} \\ \underline{\mathbf{e}}_{k|k}^{s_2}\end{pmatrix} = \begin{pmatrix}\mathbf{L}^{s_1} & 0 \\ 0 & \mathbf{L}^{s_2}\end{pmatrix}\begin{pmatrix}\underline{\mathbf{e}}_k^{s_1} \\ \underline{\mathbf{e}}_k^{s_2}\end{pmatrix} + \begin{pmatrix}\mathbf{K}^{s_1} & 0 \\ 0 & \mathbf{K}^{s_2}\end{pmatrix}\begin{pmatrix}\underline{\mathbf{v}}^{s_1} \\ \underline{\mathbf{v}}^{s_2}\end{pmatrix} \ .$$

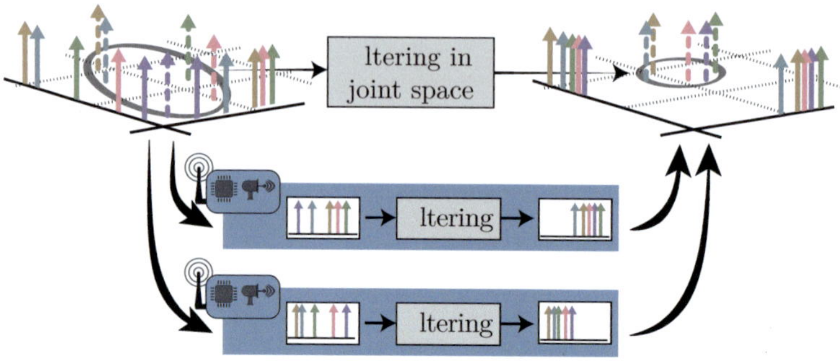

Figure 3.3: The decomposition of the joint space filtering operation operating on two-dimensional samples into two independent sensor filter steps operating on one-dimensional samples.

Hence, in order to obtain a sample representation of the joint space error, the corresponding block-diagonal matrix transformation must be applied to the augmented coordinate lists and samples for the joint space error must be generated.

Note that matrices $\mathbf{L}^s$ affect only local errors. Therefore, the transformations of the sample sets can be applied independently of each other. This also pertains to the adding of noise to the samples. As the noise terms $\underline{\mathbf{v}}^{s_1}$ and $\underline{\mathbf{v}}^{s_2}$ are independent in systems according to Definition 2.1, the random generation of samples from the joint noise term $\begin{pmatrix}\underline{\mathbf{v}}^{s_1} & \underline{\mathbf{v}}^{s_2}\end{pmatrix}^\top$ can be decomposed to the sensors by means of independent samplers. This is important as in sensor networks without global model knowledge sensors do not have access to noise distributions and transformations of remote nodes.

In particular, different samples are not combined in the filter operation due to the independence of the sample processing from the gains $\mathbf{L}^s$. Thus, the indexing of samples can be maintained and samples can be construed as independently generated. The proposed filter processing is depicted in Figure 3.3. ∎

Obviously, the scheme of Example 3.2 generalizes to more than two nodes and multivariate systems. When samples are exchanged in the fusion operation along with the estimates, (cross-)covariance matrices can be estimated. More generally, the idea pursued in this section is to **decompose the processing** of samples that describe the joint space estimation error into local sensor operations, i.e., each sensor processes sub-vectors of the joint space samples. In contrast to a purely local processing, the generation of samples is adapted to that of remote nodes, which, in particular, necessitates to generate identical samples for the same (process) noise terms at different nodes and to maintain an assignment between samples from different sets.

The proposed technique enables the estimation of covariances from local sample lists while the augmentation of sample lists also permits estimating cross-covariance matrices. This section is devoted to the presentation of the basic scheme and to the optimization of the processing for finite sample lists. The following issues are addressed:

- A **sampling policy** that enables distributed processing of joint samples in linear systems is proposed and utilized to derive an algorithm for consistent sample-based estimation of (cross-)covariance matrices.
- Covariance estimators are discussed that minimize the uncertainty for a **finite number of samples**.
- The **sampling distribution that exhibits optimality** for estimating covariances in the proposed scheme is derived.
- Methods for the **regularization** of the proposed covariance estimation scheme are discussed.

The result of this section constitutes an algorithm for sample-based covariance estimation that is asymptotically exact. The local sample processing is carried out without access to transformations at remote nodes and without global model knowledge. In particular, computational and communication effort of the algorithm is not affected by the number of emerging noise terms.

Estimated covariances can be used as a quality measure in subsequent applications, e.g., to obtain feedback matrices in control applications. Indeed, it is also feasible to use estimated (cross-)covariance matrices for the optimization of local estimators, i.e., in filter and fusion gain calculations. However, then, the gains $\mathbf{L}^s$ and $\mathbf{F}^s$ are functions of the samples so that the actual transformation of the samples with the gains in filter and fusion operations inevitably constitutes a nonlinear transformation of the samples that induces a bias in the covariance estimation for finite sample sets. Thus, the concurrent covariance estimation and gain optimization is suboptimal when the noise samples are generated independently of each other. Still, the scheme can serve as the basis for a meaningful and adjustable sensor network estimator as demonstrated in subsequent evaluations.

3.1.1 Asymptotic Properties

In the following, the basic scheme for the recursive processing of samples is presented that enables the estimation of (cross-)covariance matrices in sensor networks. The algorithm has been proposed in [186]. It consists of a sampling policy for noise distributions and transformation rules for the sample lists.

In a first instance, the estimation error in linear systems in analyzed. For this purpose, consider a linear estimator $\hat{\underline{\mathbf{x}}}$ as specified in Algorithm 2.1. The estimation error ensues from the equation $\underline{\mathbf{e}} = \hat{\underline{\mathbf{x}}} - \underline{\mathbf{x}}$ and is recursively defined as follows: at **initialization**, the estimation error is given by the error of the initial estimate, i.e., $\underline{\mathbf{e}}_0 = \hat{\underline{\mathbf{x}}}_0 - \underline{\mathbf{x}}_0$. In the **prediction** step, it follows from (2.8) that

$$\underline{\mathbf{e}}_{k+1} = \mathbf{A}\hat{\underline{\mathbf{x}}}_k - \mathbf{A}\underline{\mathbf{x}}_k - \underline{\mathbf{w}}_k = \mathbf{A}\underline{\mathbf{e}}_k - \underline{\mathbf{w}}_k \ . \quad (3.1)$$

The **filtering** of measurements with $\mathbf{K}$ and $\mathbf{L} = \mathbf{I} - \mathbf{KH}$ leads to

$$\underline{\mathbf{e}}_{k|k} = (\mathbf{I} - \mathbf{KH})\hat{\underline{\mathbf{x}}}_k + \mathbf{K}\underline{\mathbf{z}}_k - \underline{\mathbf{x}}_k = (\mathbf{I} - \mathbf{KH})\underline{\mathbf{e}}_k + \mathbf{K}\underline{\mathbf{v}}_k \ . \tag{3.2}$$

In the **fusion**, it follows with $\sum_{s\in\mathcal{S}} \mathbf{F}^s = \mathbf{I}$ that

$$\underline{\mathbf{e}} = \sum_{s\in\mathcal{S}} \mathbf{F}^s \hat{\underline{\mathbf{x}}}^s - \underline{\mathbf{x}} = \sum_{s\in\mathcal{S}} \mathbf{F}^s \hat{\underline{\mathbf{x}}}^s - \sum_{s\in\mathcal{S}} \mathbf{F}^s \underline{\mathbf{x}} = \sum_{s\in\mathcal{S}} \mathbf{F}^s \underline{\mathbf{e}}^s \ . \tag{3.3}$$

By "rolling out" the recursive calculations, it becomes obvious that the estimation error is a **linear combination of independent noise terms**. Consider for example the subsequent calculation of filter and prediction steps that is given by

$$\underline{\mathbf{e}}_{k+1} = \mathbf{A}(\mathbf{I} - \mathbf{KH})\underline{\mathbf{e}}_k + \mathbf{AK}\underline{\mathbf{v}}_k - \underline{\mathbf{w}}_k \ . \tag{3.4}$$

Apparently, the predicted error is the linearly transformed sum of the error of the previous time step and two noise terms. Repeating this process with $\underline{\mathbf{e}}_k$, the error process at $\underline{\mathbf{e}}_{k+1}$ is given as the linear combination of $\underline{\mathbf{e}}_{k-1}$ and four noise terms. Notably, the transformations of the error terms adapt to the estimator processing such that for example in (3.4), $\underline{\mathbf{v}}$ is transformed with the matrix $\mathbf{T} = \mathbf{AK}$.

Hence, in order to achieve a sample representation of estimation errors in linear systems, it is sufficient to generate samples from individual noise terms and to transform them according to the estimator processing.

3.1.1 - a Individual Noise Terms

Let $\underline{\psi}$ denote one of the noise terms that emerge in the estimation, i.e., either initial, process, or measurement noise. As in the LMMSE theory, the estimation error is sufficiently specified by its covariance, samples do not need to represent the (entire) true error distribution but only a random variable that exhibits the same mean and covariance. The following definition introduces a concept to substitute the underlying true noise distribution by a sampling distribution with the same second moment.

Definition 3.1 *A random variable* $\underline{\phi}$ ***characterizes*** *another random variable* $\underline{\psi}$ *(up to the second moment) when* $\mathrm{E}\{\underline{\psi}\} = \mathrm{E}\{\underline{\phi}\}$ *and* $\mathbf{Cov}\left(\underline{\psi}\right) = \mathbf{Cov}\left(\underline{\phi}\right)$.

Although the characterization of noise terms according to Definition 3.1 is not necessary for the derivation of asymptotic properties of the covariance estimator, it already generalizes the theory to noise terms with finite second moment from which sampling is not possible. Moreover, thanks to the characterization, the sampling distribution can be construed as a parameter of the algorithm, which is subsequently optimized in Section 3.1.2.

Note also that the sample-based covariance estimation is derived under the premise that only two moments are considered. This makes perfect sense for linear estimation, but can be too simple for nonlinear systems or in the presence of multi-modal densities. Indeed, Definition 3.1 can be easily generalized to higher moments, e.g., a sampling distribution may be admitted as substitute for a noise distribution only if other moments are matched as well.

In a next step, consider an arbitrary **noise sampler** that independently draws samples from a distribution $\underline{\phi}$ and let $\Phi \sim \underline{\phi}$ with $\Phi = (\underline{\phi}_1, \ldots, \underline{\phi}_d)$ denote a list of noise samples that are independently generated according to the distribution $\underline{\phi}$. For the sake of simplicity, sample lists are assumed to contain the same number of samples d. With $\mathbf{T} \in \mathbb{R}^{n_x \times n_x}$, the shorthand notations

$$\mathbf{T}\Phi = (\mathbf{T}\underline{\phi}_1, \ldots \mathbf{T}\underline{\phi}_d)$$

and

$$\Phi^{\nu_1} + \Phi^{\nu_2} = (\underline{\phi}_1^{\nu_1} + \underline{\phi}_1^{\nu_2}, \ldots, \underline{\phi}_d^{\nu_1} + \underline{\phi}_d^{\nu_2})$$

follow directly from matrix algebra. Let Φ^{ν_1} and Φ^{ν_2} denote two sample lists that are distributed according to $\underline{\phi}^{\nu_1}$ and $\underline{\phi}^{\nu_2}$ respectively.

Then, the (cross-)covariance matrix of $\underline{\phi}^{\nu_1}$ and $\underline{\phi}^{\nu_2}$ is asymptotically obtained with any consistent covariance estimator such as the **sample covariance** that is given by

$$\hat{\mathbf{C}}^{sc}(\Phi^{\nu_1}, \Phi^{\nu_2}) = \frac{1}{d-1} \sum_{i=1}^{d} (\underline{\phi}_i^{\nu_1} - \bar{\underline{\phi}}^{\nu_1})(\underline{\phi}_i^{\nu_2} - \bar{\underline{\phi}}^{\nu_2})^\top , \tag{3.5}$$

with sample mean $\bar{\underline{\phi}}^{\nu} = \frac{1}{d} \sum_{i=1}^{d} \underline{\phi}_i^{\nu}$. The correction of the factor $\frac{1}{d}$ in (3.5) by one is called Bessel correction. It guarantees the unbiasedness of the estimator. Properties of the sample covariance, especially regarding the minimum number of samples to achieve a desired precision have been studied in [165]. When the distribution of the samples is known, estimators that are more efficient can be derived. For example for Gaussian distributed samples, the **population covariance** defined by

$$\hat{\mathbf{C}}^{p}(\Phi^{\nu_1}, \Phi^{\nu_2}) = \frac{1}{d} \sum_{i=1}^{d} (\underline{\phi}_i^{\nu_1} - \bar{\underline{\phi}}^{\nu_1})(\underline{\phi}_i^{\nu_2} - \bar{\underline{\phi}}^{\nu_2})^\top , \tag{3.6}$$

is the maximum likelihood estimator. For the sake of a simple notation, let $\hat{\mathbf{C}}(\Phi) := \hat{\mathbf{C}}(\Phi, \Phi)$, and as the distribution $\underline{\phi}$ serves as a full substitute for the noise distribution $\underline{\psi}$ in terms of the covariance estimation, let $\underline{\psi} \equiv \underline{\phi}$ without loss of generality.

As discussed above, estimation errors are linear combinations of independent noise terms. Indeed, when two sample lists Φ^{ν_1} and Φ^{ν_2} are generated independently of each other, the true cross-covariance matrix is $\mathbf{0}$. Therefore, it holds

$$\lim_{d \to \infty} \hat{\mathbf{C}}(\Phi^{\nu_1}, \Phi^{\nu_2}) = \mathbf{0}$$

for consistent covariance estimators $\hat{\mathbf{C}}$. When the same noise term emerges in different estimators, e.g., the process noise, fully correlated samples must be generated at different sensors. For this purpose, synchronized noise samplers are used.

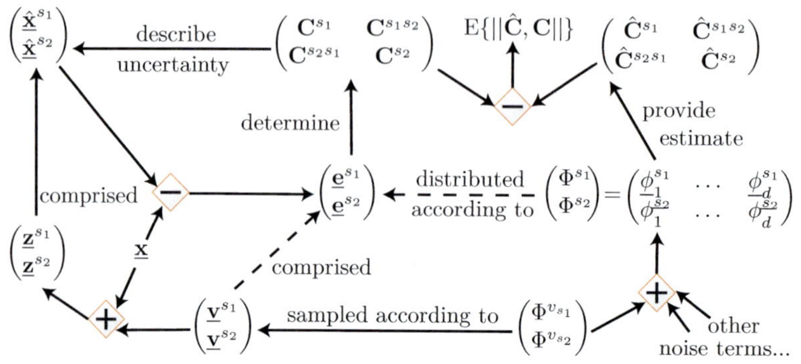

Figure 3.4: The emerging variables in sample-based covariance estimation. For simplicity, only the filtering is considered and the representation is confined to two sensors. The evaluation of the covariance estimate indicated with $\mathrm{E}\{||\hat{\mathbf{C}}, \mathbf{C}||\}$ becomes relevant in the context of finite sample lists.

Definition 3.2 *Two noise samplers that generate noise lists $\Phi^{\nu} \sim \underline{\phi}$, $\nu \in \{\nu_1, \nu_2\}$ are (second moment)* ***$\underline{\phi}$-synchronized****, if*

$$\lim_{d \to \infty} \hat{\mathbf{C}}(\Phi^{\nu_1}, \Phi^{\nu_2}) = \mathbf{Cov}\left(\underline{\phi}\right)$$

for consistent estimators $\hat{\mathbf{C}}$.

Note that synchronized noise samplers can be implemented in sensor networks without establishing a connection between nodes. For this purpose, let the sensors have access to local **pseudorandom number generators** that can be initialized with a seeding value. When, a network-wide known variable such as the current time step, a synchronized value, or the like is used to initialize the pseudorandom number generators, the same sequence of numbers and, consequently, the same samples are generated.

3.1.1 - b Reconstruction and Recursive Processing

The idea of the basic sample-based covariance estimator is to recursively process a sample list $\Phi^s = (\underline{\phi}_1^s, \ldots, \underline{\phi}_d^s)$ at each sensor that describes the estimation error $\underline{\mathbf{e}}^s$ and stores linear dependencies to other estimators. For this purpose, the evolution of the estimation error in (3.1) to (3.3) must be reflected by corresponding transformations of the sample list. An overview over the considered variables and their relation is given in Figure 3.4

According to Definition 2.1, measurement and process noise terms are independent from all other emerging noise terms. The only exception constitutes the common process noise that is modeled at all nodes and implies full correlation. For a formalization, let $\mathcal{I}$ denote a set of indices that contains unique identifiers for all noise terms $\{\underline{\mathbf{v}}_k^s, \underline{\mathbf{w}}_k, \underline{\mathbf{e}}_0^s \mid s \in \mathcal{S}\ , k \in \mathcal{K}\}$. Then, a sum representation of the estimation error is given by

$$\underline{\mathbf{e}}^s = \sum_{i \in \mathcal{I}^s} \mathbf{T}_i^s \underline{\psi}_i \ , \tag{3.7}$$

where $\mathcal{I}^s \subseteq \mathcal{I}$ contains the indices of noise terms $\underline{\psi}_i$ that affect the error process $\underline{\mathbf{e}}^s$ and $\mathbf{T}_i^s$ denote the corresponding matrix transformations. Note that due to the fusion of estimates, remote noise terms $\{\underline{\mathbf{v}}_k^{\tilde{s}}\}_{k \in \mathcal{K}, \tilde{s} \neq s}$ and $\{\underline{\mathbf{e}}_0^{\tilde{s}}\}_{\tilde{s} \neq s}$ are entailed in $\mathcal{I}^s$ as well. As argued above, sample lists are supposed to satisfy the following definition.

Definition 3.3 *The tuple* $(\hat{\underline{\mathbf{x}}}^s, \Phi^s)$ *is a* ***valid sample representation*** *(of the state* $\underline{\mathbf{x}}$*), if* $\Phi^s \sim \sum_{i \in \mathcal{I}^s} \mathbf{T}_i^s \underline{\phi}_i = \hat{\underline{\mathbf{x}}}^s - \underline{\mathbf{x}}$ *and noise terms* $\underline{\phi}_i$ *are sampled according to the following policy:*

- *Process noise is sampled with* $\underline{\phi}_i$*-synchronized noise samplers,*
- *Measurement noise is sampled with independent noise samplers.*

It is a direct consequence of $\Phi \sim \underline{\hat{\mathbf{x}}} - \underline{\mathbf{x}}$ that it holds

$$\lim_{d \to \infty} \hat{\mathbf{C}}(\Phi) = \mathrm{E}\{(\underline{\hat{\mathbf{x}}} - \underline{\mathbf{x}})^2\} ,$$

for consistent covariance estimators $\hat{\mathbf{C}}$. A recursive processing of tuples $(\underline{\hat{\mathbf{x}}}, \Phi)$ that ensures Definition 3.3 is easily obtained. At initialization, a list of samples generated by an arbitrary noise sampler that reflects the covariance $\mathbf{C}_0$ satisfies the claim. In prediction, filtering, and fusion steps, samples are transformed in conformity with the estimation error as specified in Algorithm 3.2.

Algorithm 3.2 Sensor Processing of the Sample-based Cov. Est.

1: Initialization: $(\underline{\hat{\mathbf{x}}}, \Phi) = (\underline{\hat{\mathbf{x}}}_0, \Phi_0)$, $\Phi_0 \sim \underline{\hat{\mathbf{x}}}_0 - \underline{\mathbf{x}}_0$
2: **for** $k = 1; k \in \mathcal{K}; k = k + 1$ **do**
3: Prediction: $(\mathbf{A}\underline{\hat{\mathbf{x}}}, \mathbf{A}\Phi + \Phi^w)$, $\Phi^w \sim \underline{\mathbf{w}}_k$ ($\underline{\mathbf{w}}_k$-synchronized generation)
4: Filtering: $(\mathbf{L}\underline{\hat{\mathbf{x}}} + \mathbf{K}\underline{\mathbf{z}}, \mathbf{L}\Phi + \mathbf{K}\Phi^v)$, $\Phi^v \sim \underline{\mathbf{v}}_k$ (independent)
5: Fusion: $(\sum_{s \in \mathcal{S}} \mathbf{F}^s \underline{\hat{\mathbf{x}}}^s, \sum_{s \in \mathcal{S}} \mathbf{F}^s \Phi^s)$
6: **end for**

Note that the samples are generated and transformed based on local sensor knowledge. In particular, computational and communication effort is not determined by the number of considered noise terms. The result is summarized in the following theorem.

Theorem 3.1 *Let the models be linear and let all nodes employ the linear estimator from Algorithm 3.2. Then, it holds*

$$\lim_{d \to \infty} \hat{\mathbf{C}}(\Phi^s, \Phi^{\tilde{s}}) = \mathbf{C}^{s\tilde{s}}, \ \forall s, \tilde{s} \in \mathcal{S},$$

for consistent covariance estimators $\hat{\mathbf{C}}$.

PROOF. See Appendix B. □

A detailed evaluation of the technique is provided in [186]. The next example serves as a short summary of these results.

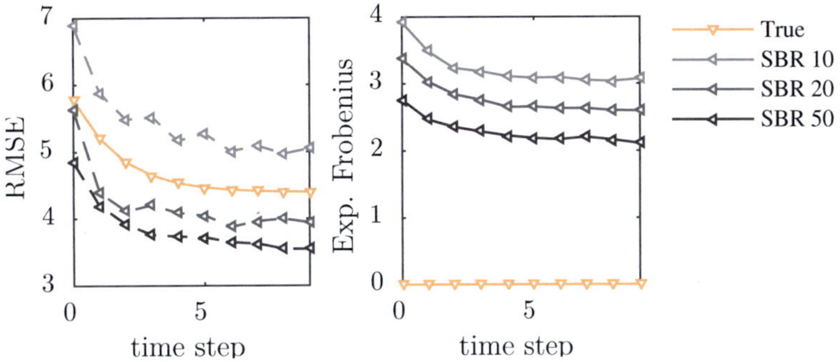

Figure 3.5: In the left figure, the RMSE of the estimate $\hat{\underline{\mathbf{x}}}^{s_3}$ and the corresponding estimated values for one run are depicted. In the right figure, the expected Frobenius norm of the covariance estimate in 100 Monte Carlo runs is given.

Example 3.3: Evaluation of Sample-set Sizes

Consider a sensor network consisting of three nodes, where only nodes s_1 and s_2 are equipped with sensor devices. A bidirectional communication is established between nodes s_1 and s_3 as well as between nodes s_2 and s_3 at each time step. The state is two-dimensional with transition matrix $\mathbf{A} = \mathbf{I}$ and zero-mean Gaussian noise with $\mathbf{Q} = \mathbf{I}$. The sensors observe different dimensions of the state according to $\mathbf{H}^{s_1} = (1 \quad 0)$ and $\mathbf{H}^{s_2} = (0 \quad 1)$, where $\{\underline{\mathbf{v}}_k^s\}_{k \in \mathcal{K}}$ is Gaussian distributed with $\mathbf{R}^s = 20$, $s \in \{s_1, s_2\}$.

The sensors are initialized with independent estimates defined by $\mathbf{C}^s = 100 \cdot \mathbf{I}$ and employ constant gain filters with $\mathbf{K} = 0.5 \cdot (\ 1 \ \ 1\ )^\top$, $\mathbf{L}^s = \mathbf{I} - \mathbf{K}\mathbf{H}^s$. In the fusion, the average of the received values is calculated. The covariances are obtained with Algorithm 3.2 and noise samples are generated from Gaussian distributions. The precision of the covariance estimates is assessed by means of the expected Frobenius norm $\mathrm{E}\left\{\left(\operatorname{tr}\left\{(\hat{\mathbf{C}} - \mathbf{C})^2\right\}\right)^{\frac{1}{2}}\right\}$.

The results of the sample-based covariance estimation for sample list sizes 10, 20, and 50 are depicted in Figure 3.5. Note that the Frobenius norm takes into account the off-diagonal entries of the covariance as well so that its absolute value appears high compared to the RMSE, which is a function of the diagonal elements only. While, in the considered run, the estimated RMSE approximately matches the true RMSE for all sample list sizes, an analysis of the Frobenius norm over 100 Monte Carlo runs shows that the expected precision is increased with additional samples. ■

3.1.2 Finite Sample Lists

As it becomes apparent from Example 3.3, the error of the covariance estimation for finite sample lists is not negligible. Indeed, in real sensor networks, computational resources are limited so that only finite sample lists can be processed. As the sample-based covariance estimation allows the characterization of noise distributions, it is tempting to use deterministic sampling methods, which, in general, provide more precise covariance estimates than stochastic sampling techniques.

However, note that the proposed processing is a decomposition of a joint space sampling scheme so that the rigorous application of nonlinear filter methods, e.g., of the unscented KF, would require calculating the Cholesky decomposition of the joint covariance matrix, which is not available to individual sensors. Even though deterministic sampling of local noise terms is possible, it must be ensured that the samples are generated independently from all other noise terms as otherwise correlations emerge in the sample covariance for independent noise terms and distort the results. Indeed, the desired independence can hardly be achieved for deterministic sampling methods, as it requires the pairwise orthogonality between arbitrary many sample sets, which, in turn, necessitates an infinite-dimensional sample space.

Nevertheless, the covariance estimator and the distribution for the characterization of noise terms can be optimized for finite sample sizes according to a criterion such as the largest eigenvalue, the ℓ_1-norm, or the Frobenius norm [17,102]. In the following, the optimal parameters for the expected Frobenius norm, which is defined as

$$\mathrm{E}\{||\hat{\mathbf{C}} - \mathbf{C}||_F\} := \mathrm{E}\left\{ \mathrm{tr}\left\{(\hat{\mathbf{C}} - \mathbf{C})^2\right\} \right\} = \mathrm{tr}\left\{\mathbf{Cov}\left(\hat{\mathbf{C}}\right)\right\} , \tag{3.8}$$

are derived. Let $\mathbf{V}$ denote the eigenvector and $\mathbf{E}$ the (diagonal) eigenvalue matrix of $\mathbf{C}$ such that $\mathbf{C} = \mathbf{VE}(\mathbf{V})^\top$. As $\mathbf{V}$ is orthogonal for symmetric matrices $\mathbf{C}$, it follows with the cyclic property of the trace

$$\begin{aligned}
\mathrm{tr}\left\{(\hat{\mathbf{C}} - \mathbf{C})^2\right\} &= \mathrm{tr}\left\{(\mathbf{V}(\hat{\mathbf{E}} - \mathbf{E})(\mathbf{V})^\top)^2\right\} \\
&= \mathrm{tr}\left\{\mathbf{V}(\hat{\mathbf{E}} - \mathbf{E})^2(\mathbf{V})^\top\right\} \\
&= \mathrm{tr}\left\{(\mathbf{V})^\top\mathbf{V}(\hat{\mathbf{E}} - \mathbf{E})^2\right\} \\
&= \mathrm{tr}\left\{(\hat{\mathbf{E}} - \mathbf{E})^2\right\} ,
\end{aligned}$$

where $\hat{\mathbf{E}}$, in general, denotes a fully occupied random matrix. In the following, only unbiased sampling schemes with $\mathrm{E}\{\hat{\mathbf{E}}\} = \mathbf{E}$ are considered so that (3.8) simplifies to

$$\mathrm{E}\{||\hat{\mathbf{C}} - \mathbf{C}||_F\} = \mathrm{tr}\left\{\mathrm{E}\{\hat{\mathbf{E}}^2\} - \mathbf{E}^2\right\} .$$

If no additional information about $\mathbf{E}$ is available and no regularization is applied, the sample covariance is the standard choice for finite sample sizes. It can be proven that the sample covariance is unbiased and efficient when the covariance space is viewed as an extrinsic cone [156], i.e., when the particular structure of covariances imposed by the positive definite property is neglected. However, when the distribution of the samples is set, other estimators may be more efficient.

It is for example well known that the population covariance (3.6) is the maximum likelihood estimator when samples are Gaussian distributed.

The sample-based covariance estimation considered in this section benefits from two simplifications compared to the general covariance estimation problem. First, the sampling technique can be chosen freely as long as the distribution has finite second central moment. Second, the mean of the noise samples is known to be zero. This permits, in particular, the utilization of the **natural covariance estimator**

$$\hat{\mathbf{C}}^{na}(\Phi^{\nu_1}, \Phi^{\nu_2}) = \frac{1}{d}\sum_{i=1}^{d}(\underline{\phi}_i^{\nu_1} - \mathrm{E}\{\underline{\phi}^{\nu_1}\})(\underline{\phi}_i^{\nu_2} - \mathrm{E}\{\underline{\phi}^{\nu_2}\})^\top \,, \tag{3.9}$$

which exploits known means for $\Phi^\nu \sim \underline{\phi}^\nu$, $\nu \in \{\nu_1, \nu_2\}$. In basic, but lengthy, calculations it can be shown that

$$\mathbf{var}\left(\hat{\mathbf{C}}^{na}\right) \leq \mathbf{var}\left(\hat{\mathbf{C}}^{sc}\right) \,.$$

Therefore, the natural covariance estimator is used instead of the sample covariance. For the sake of a simple notation, the identifier na is omitted, i.e., the natural covariance estimator is denoted as $\hat{\mathbf{C}}$ in the sequel.

The preceding results on asymptotic properties of sample-based covariance estimation have been derived on the foundation that the sampling distributions can differ from the underlying noise distributions. More specifically, any distribution with finite second moment such as the uniform distribution, the Gaussian distribution, etc. can be used to characterize noise terms.

3.1.2 - a Optimal Scalar Sampling

In the following, the sampling distribution that minimizes the expected Frobenius norm of the natural covariance estimator is derived for scalar noise distributions $\boldsymbol{\phi}$. For scalar samples, the sample set Φ is a vector and the sample covariance $\hat{\mathbf{C}}$ is the scalar sample variance denoted as $\hat{\mathbf{v}}$. Hence, the objective is to find the sampling distribution that minimizes

$$\mathbf{var}\left(\hat{\mathbf{v}}(\Phi)\right) = \mathrm{E}\{(\hat{\mathbf{v}}(\Phi) - \mathbf{var}\left(\boldsymbol{\phi}\right))^2\} \ . \tag{3.10}$$

As specified in Section 3.1.1, samples are drawn independently of each other. This allows deriving a simple representation of (3.10) that only depends on properties of the sampling distribution.

Lemma 3.2 *Let Φ^ν, $\nu \in \{\nu_1, \nu_2\}$ denote lists of independently generated scalar samples. Then,* $\mathbf{var}\left(\hat{\mathbf{v}}(\Phi^{\nu_1}, \Phi^{\nu_2})\right)$ *equals*

$$\frac{1}{d}\left(\mathrm{E}\{(\boldsymbol{\phi}^{\nu_1} - \mathrm{E}\{\boldsymbol{\phi}^{\nu_1}\})^2 \, (\boldsymbol{\phi}^{\nu_2} - \mathrm{E}\{\boldsymbol{\phi}^{\nu_2}\})^2\} - (\mathbf{var}\left(\boldsymbol{\phi}^{\nu_1}, \boldsymbol{\phi}^{\nu_2}\right))^2\right) \ .$$

PROOF. See Appendix B. □

Hence, the precision of the variance estimator depends on the number of samples and a fourth order term. For **independently generated sample lists** $\boldsymbol{\phi}^{\nu_1}$ and $\boldsymbol{\phi}^{\nu_2}$, it holds

$$\mathrm{E}\{(\boldsymbol{\phi}^{\nu_1} - \mathrm{E}\{\boldsymbol{\phi}^{\nu_1}\})^2(\boldsymbol{\phi}^{\nu_2} - \mathrm{E}\{\boldsymbol{\phi}^{\nu_2}\})^2\} = \mathbf{var}\left(\boldsymbol{\phi}^{\nu_1}\right)\mathbf{var}\left(\boldsymbol{\phi}^{\nu_2}\right) \ . \tag{3.11}$$

In terms of choosing the optimal sampling distribution, (3.11) cannot be optimized as $\mathbf{var}\left(\boldsymbol{\phi}^{\nu}\right)$, $\nu \in \{\nu_1, \nu_2\}$ is specified by the noise distribution. However, the estimation of local variances with $\Phi^{\nu_1} = \Phi^{\nu_2}$, leads to

$$\mathbf{var}\left(\hat{\mathbf{v}}(\Phi)\right) = \frac{1}{d}\left(\mathrm{E}\{(\boldsymbol{\phi} - \mathrm{E}\{\boldsymbol{\phi}\})^4\} - \mathbf{var}\left(\boldsymbol{\phi}\right)^2\right) \ . \tag{3.12}$$

Using the definition of the **excess kurtosis**[2], i.e.,

$$\kappa = \frac{\mathrm{E}\{(\boldsymbol{\phi} - \mathrm{E}\{\boldsymbol{\phi}\})^4\}}{\mathbf{var}\left(\boldsymbol{\phi}\right)^2} - 3 \ ,$$

[2]The excess kurtosis is the normalized kurtosis so that $\kappa = 0$ for Gaussian variables.

the variance from (3.12) is transformed to

$$\mathbf{var}\left(\hat{\mathbf{v}}(\Phi)\right) = \frac{\mathbf{var}\left(\boldsymbol{\phi}\right)^2}{d}\left(\kappa + 2\right) . \tag{3.13}$$

The excess kurtosis is a distribution-dependent variable with various interpretations, e.g., it can be construed as a degree of bimodality of the density [38]. Therefore, the more the density resembles bimodal distributions, the higher is the precision in the variance estimator. In particular, it holds $\kappa \geq -2$, where $\kappa = -2$ is attended for a Bernoulli distribution with equal success and failure probability [38]. Affine transformations of samples from this distribution permit the specification of a **modified Bernoulli distribution** with

$$\Pr\left\{\boldsymbol{\phi} = -\sigma\right\} = 0.5 \text{ and } \Pr\left\{\boldsymbol{\phi} = \sigma\right\} = 0.5 , \tag{3.14}$$

with zero mean and configurable standard deviation σ. Hence, the estimator variance from (3.12) is minimized when the samples are distributed according to (3.14). Considering a noise term $\boldsymbol{\phi}$ with $\mathrm{E}\{\boldsymbol{\phi}\} = 0$ and $\mathrm{E}\{\boldsymbol{\phi}^2\} = \mathbf{var}\left(\boldsymbol{\phi}\right)$, the modified Bernoulli distribution with $\sigma = \sqrt{\mathbf{var}\left(\boldsymbol{\phi}\right)}$ characterizes the noise term $\boldsymbol{\phi}$. As $\mathrm{E}\{\boldsymbol{\phi}^2\} = \sigma^2 = \boldsymbol{\phi}^2$ is deterministic, sample-based variance estimators with Bernoulli distributed samples have uncertainty zero.

In the context of linear estimation, independent and fully dependent noise terms emerge, and are sampled and transformed according to the estimator processing. Let the estimation error be given by $\mathbf{e}_k^s = \sum_{i \in \mathcal{I}^s} \mathbf{T}_i^s \boldsymbol{\phi}_i$ (3.7). Then, sample lists are linearly transformed noise sample lists according to

$$\Phi^s = \sum_{i \in \mathcal{I}^s} \mathbf{T}_i^s \Phi_i^\nu ,$$

where $\Phi_i^\nu \sim \boldsymbol{\phi}_i$. Therefore, the expected Frobenius norm of the variance estimator $\hat{\mathbf{v}}(\Phi^s)$ is given by the sum of terms of the form

$$(\mathbf{T}_i^s)^2 (\mathbf{T}_{\tilde{i}}^s)^2 \, \mathrm{E}\{(\Phi_i^\nu)^2 (\Phi_{\tilde{i}}^\nu)^2\} \text{ with } i, \tilde{i} \in \mathcal{I}^s .$$

As in the considered linear system, noise terms are independent, the terms for $i \neq \tilde{i}$ are specified by (3.11). For terms with $i = \tilde{i}$, the Bernoulli distribution minimizes the variance. In summary, the following statement is obtained.

Theorem 3.3 *Let* $(\hat{\mathbf{x}}^s, \Phi^s)$ *denote valid sample representations according to Definition 3.3,* $s \in \mathcal{S}$*. Then,* $\mathbf{var}(\hat{\mathbf{v}}(\Phi^{s_1}, \Phi^{s_2}))$ *is minimized when noise samples are generated from the modified Bernoulli distributions with* $\sigma = \sqrt{\mathbf{var}(\phi)}$ *as defined in* (3.14).

It is worth mentioning that the sample representation induces an estimation error even for uncorrelated noise terms. In fact, every time new noise samples are generated and added, co-variance terms (3.11) emerge in the expected Frobenius norm for all noise terms that are already comprised in the estimation error.

In particular, the sample-based variance estimation is affected by stochastic errors even if samples are generated from the Bernoulli distribution. Still, for individual noise terms, the variance estimator with Bernoulli distributed samples is error-free.

Samples of the estimation error are in general not Bernoulli distributed. Indeed, as long as the individual noise samples are Bernoulli distributed the expected Frobenius norm from Lemma 3.2 can still be solved in closed form.

Theorem 3.4 *Let* $(\underline{\hat{\mathbf{x}}}, \Phi)$ *denote a valid sample representation of the state* $\underline{\mathbf{x}}$ *with Bernoulli distributed noise samples. Let furthermore* $\mathbf{T}_i$ *denote the transformation factors and* ϕ_i^ν *the noise terms from the error decomposition* (3.7) *and* $\mathbf{C}$ *denote the variance of* $\hat{\mathbf{x}}$*. It holds*

$$\mathbf{var}(\hat{\mathbf{v}}(\Phi)) = \frac{2}{d}\left((\mathbf{var}(\mathbf{C}))^2 - \sum_{i \in \mathcal{I}} (\mathbf{T}_i)^4 (\mathbf{var}(\phi_i^\nu))^2\right) . \qquad (3.15)$$

A proof of Theorem 3.4 is provided in Appendix B. A consequence of the statement is that the variance of the sample-based estimator

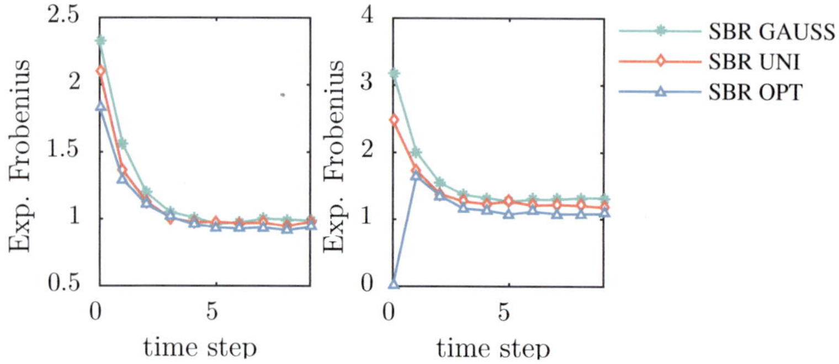

Figure 3.6: The expected Frobenius norm of the variance estimates of sensor s_1 (left) and s_2 (right) obtained with Algorithm 3.2 for Gaussian (SBR GAUSS), uniform (SBR UNI), and Bernoulli (SBR OPT) distributed noise samples.

diminishes proportional to $\frac{1}{d}$. The proportionality factor is given by the square of the true variance minus deterministic local noise terms. As a matter of fact, formulas for the recursive calculation of (3.15) can be easily obtained so that the tracking of the precision is possible.

Example 3.4: Optimal Distribution for Sample-based Cov. Est.

In this example, the generation of noise samples according to different distributions is examined. To this end, consider a scalar linear system with $\mathbf{A} = 1$ and $\mathbf{Q} = 1$. The state is observed by two sensors with $\mathbf{H}^s = 1$, $\mathbf{R}^s = 20$, and initial uncertainty $\mathbf{C}_0^s = 40$, $s \in \{s_1, s_2\}$. Both sensors employ constant gain filters with $\mathbf{K} = 0.5$, $\mathbf{L} = 0.5$, and the average operator as fusion operation. While sensor s_1 receives the latest estimate from sensor s_2 in each time step, sensor s_2 operates only on local data. The precision of sample-based variance estimators with 20 samples in terms of the expected Frobenius norm in 500 Monte Carlo runs is depicted in Figure 3.6.

The results differ for the noise distributions. Especially in the first time steps when the number of involved noise terms is small, the sample representation of individual noise terms determines the estimation error so that the optimization of the precision by means of the Bernoulli distribution yields significantly smaller errors. In fact, when the estimation error consists of only one noise term, as in time step 0 at sensor s_2, the variance estimate for Bernoulli distributed samples is exact. With an increasing number of noise terms, the precision of the variance estimates becomes primarily determined by the estimation errors of co-variances (3.11), which are the same for all sampling distributions. Therefore, the difference in estimation performance diminishes when time progresses. Concurrently, the absolute precision of the variance estimate improves. This is due to Theorem 3.4, which establishes a direct connection between the variance of the estimate and the precision of the variance estimator. As the variance of the estimate decreases in the considered setup, this translates into an improvement of the variance estimators.

In summary, the usage of the modified Bernoulli distribution improves the precision of sample-based variance estimators for finite sample sizes. However, other distributions achieve similar results when several noise terms are involved. ■

By means of a probabilistic argument, the optimal scalar sampling distribution for the sample-based variance estimation of Algorithm 3.2 has been derived. In particular, it could be demonstrated that by means of the Bernoulli distribution, individual noise terms can be deterministically determined.

3.1.2 - b Optimal Sampling for Multivariate Systems

For an application of the sampling technique to arbitrary systems, generalizations of Theorems 3.3 and 3.4 to multivariate systems are desired. Considering $\mathrm{E}\{||\hat{\mathbf{C}} - \mathbf{C}||_F\} = \mathrm{tr}\left\{\mathrm{E}\{\hat{\mathbf{E}}^2\} - \mathbf{E}^2\right\}$, for unbiased sampling schemes, the minimization of the expected Frobenius norm boils down to the optimal sampling of eigenvalue matrices. Indeed, the sampling of eigenvalue matrices induces errors in the off-diagonals as the matrix is estimated from (sample) vectors.

Lemma 3.5 *Let $\Phi \sim \underline{\phi}$ denote a list of samples. Then,*

$$\mathrm{tr}\left\{\mathbf{Cov}\left(\hat{\mathbf{C}}(\Phi)\right)\right\} = \frac{1}{d}\left((\underline{\mathbf{1}})^\top \breve{\mathbf{C}}\underline{\mathbf{1}} - \mathrm{tr}\left\{\mathbf{E}^2\right\}\right) , \tag{3.16}$$

where $[\breve{\mathbf{C}}]_{\rho\tilde{\rho}} = \mathrm{E}\{([\underline{\phi}]_\rho - \mathrm{E}\{[\underline{\phi}]_\rho\})^2([\underline{\phi}]_{\tilde{\rho}} - \mathrm{E}\{[\underline{\phi}]_{\tilde{\rho}}\})^2\}$.

PROOF. See Appendix B. □

Considering a noise term $\underline{\psi}$ with $\mathrm{E}\{\underline{\psi}(\underline{\psi})^\top\} = \mathbf{E}$, sampling distributions $\underline{\phi}$ that characterize $\underline{\psi}$ must satisfy

$$\mathrm{E}\{(\underline{\phi} - \mathrm{E}\{\underline{\phi}\})^2\} = \mathbf{E} .$$

Hence, it follows from Lemma 3.5 that the minimization of $(\underline{\mathbf{1}})^\top\breve{\mathbf{C}}\underline{\mathbf{1}}$ poses the main challenge. The optimal solution to this problem is given in the next statement, which is proven in Appendix B.

Theorem 3.6 *Let $\underline{\psi}$ denote a zero-mean noise term with covariance $\mathbf{VE}(\mathbf{V})^\top$, where $\mathbf{E}$ denotes the diagonal eigenvalue and $\mathbf{V}$ the corresponding eigenvector matrix. Then, $\mathrm{E}\{||\hat{\mathbf{C}}(\Phi) - \mathbf{C}||_F\}$ is minimized when the sample lists Φ are distributed according to a* ***multivariate modified Bernoulli distribution*** *with $\underline{\phi} = \mathbf{V}\underline{\tilde{\phi}}$, where $[\underline{\tilde{\phi}}]_\rho$ is independently distributed according to (3.14) with $\sigma = \sqrt{\mathbf{E}_\rho}$, $\rho \in \{1, \ldots, n_x\}$.*

A comparison of Theorem 3.3 with Theorem 3.6 reveals that the statement of the latter is weaker. While for scalar systems, the optimal distribution for the **estimation error** is specified, the statement for multivariate systems is confined to individual noise terms. This is a consequence of matrix transformations that are applied to noise terms in linear estimation. Even if the eigenvalues of all noise terms are Bernoulli distributed at a specific time step, the transformation, e.g., with the state transition matrix $\mathbf{A}$, modify the eigenvalues so that in subsequent time steps, the eigenvalues are no longer Bernoulli distributed.

It is also worth mentioning that only the diagonal entries of $\mathbf{E}$ are estimated without stochastic error. As the expected Frobenius norm comprises the product of the difference between estimated and true covariance, off-diagonal elements affect the trace in Lemma 3.5 as well. The estimation of these off-diagonal elements is comparable with the estimation of scalar variances between independently generated sample sets, which implies the error (3.11). Therefore, even the estimation of (multivariate) noise terms succumbs a stochastic error.

In summary, the error in the noise covariance estimation is minimized when the samples are generated by means of Bernoulli distributed samples. For scalar state spaces, the optimality applies also to estimation errors, where samples are linearly transformed after generation. Otherwise, the usage of the Bernoulli distribution can still be motivated by means of Theorem 3.6.

3.1.2 - c Enforcing Local Covariances

Regularization techniques can be applied to improve the precision of covariance estimators, if additional knowledge about the estimated variable is available. It is, for example, possible to penalize nonzero cross-covariance matrices between noise terms that are known to be independent of each other. In the following, a brief state-of-the-art overview over regularization techniques is given before a solution that fits to the estimation in linear systems is presented.

Pioneering work towards improving standard estimators has been proposed with the James-Stein-Estimator [82] that outperforms standard least squares estimators in terms of the MSE. The underlying idea is called **shrinkage**. It encompasses approaches that (often linearly) combine the estimate with some additional information or belief. Applied to covariance estimation, the optimal shrinkage intensity, i.e., the optimal weighting, for the identity matrix and the sample covariance with respect to the Frobenius norm has been derived [102].

Other types of regularization such as **enforcing sparsity** have been examined as well. For graphical models, conditional independence of nodes results in zero-entries in the inverse covariance. Hence, ℓ_1-penalties, so called lasso methods, applied to the inverse covariance enforce the desired sparsity of the graphical model as it is achieved, e.g., by Graphical lasso [49]. Ensuring sparsity by directly setting entries of the covariance to zero according to hard thresholds has been examined in [95]. When the ordering of the covariance elements defines some form of closeness, banding techniques are utilized to neglect covariance entries that are too far away from the off-diagonal [17, 140]. Alternatively, a less strict approach, called tapering, can be applied to shrink off-diagonal elements to zero. Convenient forms to regularize covariances [140] or its inverse [134] are regression-based methods that utilize the Cholesky decomposition. The idea is to apply a stepwise regression with increasing sample sizes, yielding a covariance decomposition on which regularization, e.g., banding, is easily realizable. A comprehensive overview of different covariance estimation methods is given in [135].

The regularization applied in the following is based on the property that local covariances of estimates can be exactly calculated in prediction (2.22) and filtering (2.24) operations. Hence, the idea is to **align the samples** to the true local covariances in order to improve the estimates of the joint covariance matrix.

Indeed, it is worth pointing out that the regularization developed in the following is only applicable when local covariances, i.e., the block diagonal of the joint covariance matrix, can be calculated. In decentralized estimation with regular fusion operations or when the models are nonlinear, only bounds or approximations for the true local covariances are computable, which, in turn, impedes the application of the proposed regularization technique.

In [186], it has been proposed to modify the noise sampling procedure in order to achieve the desired equality between estimated and true local covariance. In the sequel, a more intuitive approach is presented that introduces an additional regularization operation so that noise can be sampled without modification and the parity between estimated and true local covariance is enforced only if required. The key operation is stated next.

Proposition 3.7 *Let* $\hat{\mathbf{C}}(\Phi)$ *denote a positive definite covariance estimate obtained by means of the sample list* Φ *and let* $\mathbf{C}$ *denote a positive definite matrix. The sample list* $\tilde{\Phi} = \mathbf{T}\Phi$ *with*

$$\mathbf{T} = \sqrt{\mathbf{C}}\,\sqrt{\hat{\mathbf{C}}(\Phi)}^{-1} , \tag{3.17}$$

where $\sqrt{\cdot}$ *denotes a regular matrix square root such as the Cholesky decomposition, satisfies* $\hat{\mathbf{C}}(\tilde{\Phi}, \tilde{\Phi}) = \mathbf{C}$.

PROOF. It follows from (3.9) that $\hat{\mathbf{C}}(\mathbf{T}\Phi, \mathbf{T}\Phi) = \mathbf{T}\hat{\mathbf{C}}(\Phi)(\mathbf{T})^\top$. Therefore, for given covariances $\hat{\mathbf{C}}(\Phi)$ and $\mathbf{C}$, the challenge is to derive a transformation matrix $\mathbf{T}$ such that $\mathbf{T}\hat{\mathbf{C}}(\Phi)(\mathbf{T})^\top = \mathbf{C}$. Indeed, the claim follows with

$$\begin{aligned}\mathbf{T}\hat{\mathbf{C}}(\Phi)(\mathbf{T})^\top =& \sqrt{\mathbf{C}}\sqrt{\hat{\mathbf{C}}(\Phi)}^{-1}\sqrt{\hat{\mathbf{C}}(\Phi)}\sqrt{\hat{\mathbf{C}}(\Phi)}^{\top}\sqrt{\hat{\mathbf{C}}(\Phi)}^{-\top}(\sqrt{\mathbf{C}})^\top \\ =& \mathbf{C} \ .\end{aligned}$$

□

By means of Proposition 3.7, equality between estimated and true covariance can be restored at any time. Indeed, the estimated covariance $\hat{\mathbf{C}}(\Phi)$ is positive definite for $d \geq n_x$ independently generated samples. It is also worth mentioning that samples of synchronized samplers at different sensors are not guaranteed to be equal when Proposition 3.7 is applied. However, the asymptotic consistency of valid sample representations is not affected by the regularization as $\lim_{d\to\infty} \hat{\mathbf{C}}(\Phi) = \mathbf{C}$ so that $\mathbf{T} = \mathbf{I}$ for $d \to \infty$.

Moreover, in order to save storage and communication costs, estimates can be integrated into the sample representation Φ by enforcing the average of the samples to coincide with $\hat{\underline{\mathbf{x}}}$ and by obtaining the sample covariance with a corrected expectation vector. A more detailed examination of this idea is given in [186].

3.1.3 Sensor Network Implementation

By means of the theory for finite sample lists, an efficient technique for the estimation of (cross-)covariance matrices in sensor networks can be realized. As proposed in Algorithm 3.2, sensors omit the calculation of local covariances and instead maintain sample lists Φ that depict the estimation error of estimates. As it has been demonstrated, a recursive processing of these sample lists is feasible by applying the estimator transformations to Φ.

For the generation of samples for measurement and process noise terms, each sensor needs access to **two noise samplers**. Measurement noise samples are generated independently from other samples by using a pseudorandom number generator. The process noise of a particular time step is the same for all estimates. Hence, synchronized noise samplers that generate the same samples at all sensors are implemented by means of pseudorandom number generators that are initialized with the same seed value, e.g., the current time step.

Let $\mathbf{VE}(\mathbf{V})^\top$ denote the eigendecomposition of the noise covariance $\mathbf{C}_i^\nu$. As the uncertainty of the sample-based covariance estimation is minimized for Bernoulli distributed samples and the generation of Bernoulli samples is simple and efficient, noise sample lists Φ^ν are generated according to $\underline{\phi}_i^\nu \sim \mathbf{V}\tilde{\underline{\phi}}$ with independent $[\tilde{\underline{\phi}}]_\rho$, where

$$\begin{rcases} \Pr\left\{[\tilde{\underline{\phi}}]_\rho = -\sqrt{[\mathbf{E}]_\rho}\right\} &= 0.5 \\ \Pr\left\{[\tilde{\underline{\phi}}]_\rho = \sqrt{[\mathbf{E}]_\rho}\right\} &= 0.5 \end{rcases} , \rho \in \{1, \dots, n_x\} \ .$$

As argued in Section 3.1.2, the **natural covariance estimator** provides estimates with lower uncertainty than the sample covariance. For zero-mean noise terms, it holds $\mathrm{E}\{\underline{\mathbf{e}}\} = \underline{0}$, and (3.9) simplifies to

$$\hat{\mathbf{C}}(\Phi^{\nu_1}, \Phi^{\nu_2}) = \frac{1}{d}\Phi^{\nu_1}(\Phi^{\nu_2})^\top \ .$$

In Algorithm 3.2, the **covariance estimation** is not explicitly listed, as its application area depends on the scenario. Estimators for time-variant systems usually rely on the covariance for the optimization of filter and fusion gains so that $\hat{\mathbf{C}}$ can be used as a substitute in these operations. Then, however, dependencies between samples emerge that distort the covariance estimate as contended in the introduction of this section. Still, the scheme can yield promising results as demonstrated in the next example.

Example 3.5: Sample-based Covariance Estimation versus CI

A scalar system with $\mathbf{A} = 1.2$ and process noise variance $\mathbf{Q} = 1$ is considered. The state is observed by 10 sensors with $\mathbf{H}^s = 1$ and noise variances $\mathbf{R}^s = 12$ for $s = s_1, s_2$, $\mathbf{R}^s = 14$ for $s = s_3, s_4$, and so on up to $\mathbf{R}^s = 20$ for $s = s_9, s_{10}$. The network topology is a ring where sensors communicate with their direct neighbors and sensor s_1 is connected to sensor s_{10}. Communication is established at every time step subject to a packet loss probability of 0.25.

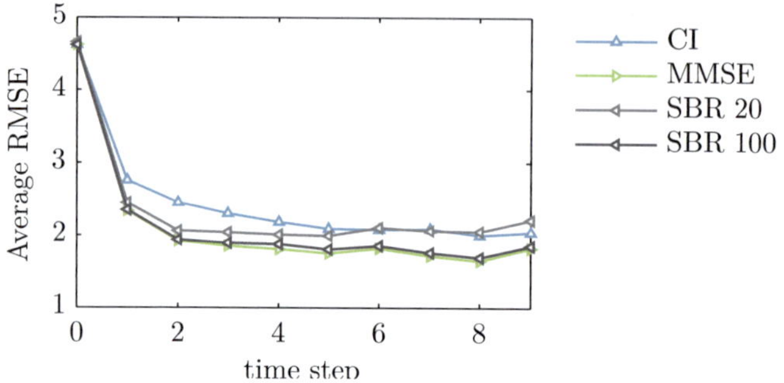

Figure 3.7: The average RMSE of all sensors at different time steps in 100 Monte Carlo runs for the compared estimators.

Overall, three processing schemes are compared. As the baseline, the local KF processing in combination with the LMMSE fusion from Theorem 2.5 is realized by means of an omniscient observer, which defines the one-time-step optimal solution to decentralized estimation. The sample-based scheme uses the same estimator but optimizes the gains by means of estimated (cross-)covariance matrices obtained with Algorithm 3.2 and Bernoulli distributed noise samples. Additionally, the estimates provided by covariance intersection [85] in combination with local KFs are compared.

As it can be seen in Figure 3.7, the sample-based covariance estimator improves with the number of processed samples. The fusion of estimates with covariance intersection does not exploit correlations. Therefore, the fusion result cannot be better than the prior estimate with the smallest bound. Consequently, the sample-based scheme provides a lower average RMSE than covariance intersection even though it does not take into account dependencies that emerge due to the concurrent gain optimization and covariance estimation. As a matter of fact, the sample-based estimation scheme with 100 samples yields almost the same RMSE as the baseline processing. ■

The precision of Algorithm 3.2 can be improved further by processing local covariances additionally to the sample lists as long as estimates are not fused. Then, the **regularization** from Section 3.1.2 - c can be used to eliminate stochastic errors in the estimation of local covariances. To this end, a regularization step according to Proposition 3.7 is carried out every time before $\hat{\mathbf{C}}$ is transmitted or calculated.

Finally, it is worth mentioning that only one-sided matrix multiplications are required to process samples. Therefore, the **computational effort** is even lower than for local KFs when the size of the sample list d is smaller than the dimension of the state. For data fusion, each sensor transmits the tuple $(\hat{\underline{\mathbf{x}}}^s, \Phi^s)$ with $n_x \times (d+1)$ elements to all neighbors. Hence, the **communication effort** increases with additional samples linearly in the state dimension. Indeed, it is possible to generalize the natural covariance estimator, e.g., by means of regression techniques, such that consistent estimates of (cross-)covariance matrices are also obtained for differing sample list sizes. This allows minimizing the communication effort by transmitting only subsets of locally held samples.

3.2 Bounding of Covariances

It is an inevitable consequence of restricted model knowledge that the exact calculation of cross-covariance matrices requires disproportionate effort and the sample-based covariance estimation provides only uncertain quality information. Hence, a computationally efficient approach that enables reliable covariance estimation with quality guarantees is still not available.

This is aggravated by practical problems often neglected in LMMSE theory. Namely, due to modeling errors, unknown perturbations, and other influences, covariances of initial estimates and those of noise terms emerging in state and measurement models are often imprecisely known. A variety of techniques to cope with such systematic uncertainties has been proposed [119, 162]. For example, if the systematic uncertainty is the consequence of the linearization by means of a Taylor series [168], the error in the model can be inferred from the underlying function. A more general approach is to handle systematic uncertainties with set theoretic methods [121]. This, however, may lead to conservative results and increases the complexity in the considered stochastic framework, as additional set variables must be processed.

A purely stochastic approach for the handling of model errors and unknown dependencies is to inflate emerging covariances artificially such that a conservative bound according to the next definition is obtained.

Definition 3.4 *Let* $\mathbf{C}$ *denote the covariance of a random variable. A* ***(covariance) bound*** *is a symmetric matrix* $\mathbf{P}$ *that satisfies* $\mathbf{P} \geq \mathbf{C}$ *in the positive semi-definite sense. Tuples of estimates* $\hat{\underline{\mathbf{x}}}$ *and bounds* $\mathbf{P}$ *with* $\mathbf{P} \geq \mathrm{E}\{(\hat{\underline{\mathbf{x}}} - \underline{\mathbf{x}})^2\}$ *are termed* ***consistent****.*

Covariance bounds can be understood as generalizations of covariances. Instead of enforcing the equality $\mathbf{C} = \mathrm{E}\{(\hat{\underline{\mathbf{x}}} - \underline{\mathbf{x}})^2\}$, they are supposed to satisfy the inequality $\mathbf{P} \geq \mathrm{E}\{(\hat{\underline{\mathbf{x}}} - \underline{\mathbf{x}})^2\}$. Bounds also integrate perfectly into linear estimation theory since they can be used as **substitutes of covariances**.

An analysis of the formulas for the calculation of covariances in the different processing steps shows that all formulas (2.20), (2.22) and (2.24) feature the same structure $\mathbf{C} = \tilde{\mathbf{T}}\tilde{\mathbf{C}}(\tilde{\mathbf{T}})^\top + \breve{\mathbf{T}}\breve{\mathbf{C}}(\breve{\mathbf{T}})^\top$ with covariances $\tilde{\mathbf{C}}$, $\breve{\mathbf{C}}$ and linear transformation matrices $\tilde{\mathbf{T}}$, $\breve{\mathbf{T}}$. Therefore, with bounds $\tilde{\mathbf{P}} \geq \tilde{\mathbf{C}}$, $\breve{\mathbf{P}} \geq \breve{\mathbf{C}}$, it follows from Observation 7.7.2 in [76] that the true covariance after the operation is bounded as well, i.e.,

$$\mathbf{C} = \tilde{\mathbf{T}}\tilde{\mathbf{C}}(\tilde{\mathbf{T}})^\top + \breve{\mathbf{T}}\breve{\mathbf{C}}(\breve{\mathbf{T}})^\top \leq \tilde{\mathbf{T}}\tilde{\mathbf{P}}(\tilde{\mathbf{T}})^\top + \breve{\mathbf{T}}\breve{\mathbf{P}}(\breve{\mathbf{T}})^\top = \mathbf{P} \,. \qquad (3.18)$$

Analogous considerations permit the substitution of covariances in the fusion operation, which, in summary, reveals the following observation.

Observation 3.8 *A bound on the true covariance of linear estimators is obtained by replacing covariances with bounds in the calculations, i.e., bounds remain valid when they are processed as substitutes of covariances.*

Therefore, the equation $\mathbf{P} \geq \mathbf{C}$ is still fulfilled when initial uncertainties or noise covariances are overestimated. On the other hand, bounds describe the underlying covariance exactly as long as no approximations due to the fusion of estimates or uncertainties in the model are applied.

Now, consider filter and fusion operations of linear estimators. In the LMMSE framework, gains, e.g., in the KF, are supposed to optimize the covariances of estimates. However, in the light of the above, these covariances are unknown and are substituted by bounds. Hence, the idea pursued in this section is to **minimize covariance bounds**. This is a meaningful generalization of the mean squared error theory as it corresponds to the LMMSE when bounds are exact and minimizes the worst-case uncertainty otherwise. Indeed, bounds can then be employed as substitutes for the unknown covariances in the formulas such that filter and fusion gains with substituted covariances minimize the bounds instead of the true covariances. The relation between bounds and covariances is depicted in Figure 3.8.

In order to provide a bound on the true covariance in the fusion under unknown cross-covariance matrices, the **set of all possible covariances** of the fused estimate must be considered. Indeed, this set is bounded, as the set of possible correlations (and cross-covariance matrices) is restricted [52, 84, 114]. Now, several challenges arise. In a first step, bounds for given gains and covariances need to be derived. Then, the optimal bound subject to a meaningful optimization criterion $\mathcal{J}\{\cdot\}$ is to be calculated. Finally, the question arises whether

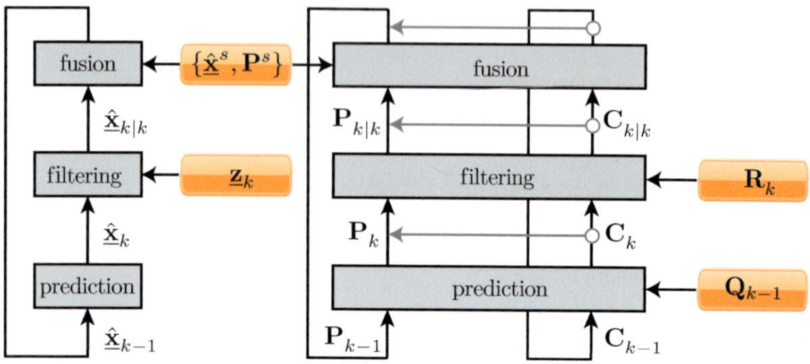

Figure 3.8: The processing of estimates, covariances, and bounds.

there are gains that constrain the set of possible covariances in a way that an especially small bound is obtained. The latter challenge is given by

$$\arg\min_{\mathbf{F}^s, s \in \mathcal{S}} \mathcal{J}\left\{\mathbf{P}\right\} \text{ with } \mathbf{P} \geq \mathrm{E}\left\{\left(\sum_{s \in \mathcal{S}} \mathbf{F}^s \hat{\underline{\mathbf{x}}}^s - \underline{\mathbf{x}}\right)^2\right\} \tag{3.19}$$

for unknown joint covariances matrix $\bar{\mathbf{C}}$. The derivation of such bounds $\mathbf{P}$ has been examined in different areas of literature. Historically, **set theory** was investigated first. Therefore, some of the results concerned with covariance bounding are special versions of statements from ellipsoidal set theory or rely heavily on them. Pioneering work pertaining to the bounding of ellipsoids has, for example, been carried out in [146] and [101].

In the context of stochastic estimation, bounds are tried to achieve using the maximum eigenvalues of prior covariances. To this end, the **Löwner-John Ellipsoid**, i.e., the largest ellipsoid contained in the intersection of prior covariance ellipsoids, is calculated to account for a bound on the mutual information. The idea proposed in [15] is to use the Löwner-John Ellipsoid to calculate a maximum covariance that is supposed to serve as a bound. A more rigorous motivation and exploitation of the corresponding theory is provided in the derivation of ellipsoidal intersection [149–151].

However, in simulations it has turned out that ellipsoidal intersection and the technique from [15] are not consistent according to Definition 3.4. This is due to potential correlations between the maximum mutual error and sensor-specific errors, which are neglected in the derivations of the algorithms.

The derivation of suitable gains for the fusion under unknown correlations has been examined in [180] by considering all possible cross-covariance matrices and the corresponding LMMSE fusion formulas. Premised on the assumption that cross-covariance matrices are uniformly distributed, optimized fusion gains are obtained.

Indeed, the most popular technique for the fusion under completely unknown correlations that guarantees consistent estimates is **covariance intersection** (CI) [85,87–89]. However, when cross-covariance matrices are completely unknown, a large set of possible outcomes must be considered to obtain a bound. Therefore, the objective of this section is to reconstruct cross-covariance matrices partially and to exploit this partial knowledge in the fusion of estimates. The key elements are as follows:

- CI is proven to provide the **smallest bound** for the fusion of two estimates under completely unknown correlations.
- A generalization of CI that takes into account **partially known cross-covariance matrices** is derived.
- Different **covariance representations** based on noise decompositions for linear systems are proposed.
- A **combination of reconstruction and bounding** techniques with tunable precision is derived.

The results constitute important properties of CI that can help to establish the procedure in theoretical and practical works concerned with the fusion of estimates. In particular, the generalized fusion method that allows including partially known correlations provides additional flexibility in designing efficient sensor network estimators. In combination with the techniques for the reconstruction of cross-covariance matrices, an efficient fusion algorithm for sensor network estimation is obtained.

3.2.1 Covariance Intersection

The idea brought forward by CI is to inflate covariances by means of scalar factors in order to account for all possible cross-covariance matrices in the fusion of estimates. Originally, it has been proposed for two sensors [85] by means of the fusion formulas

$$\begin{aligned}\underline{\hat{\mathbf{x}}}^{\omega} &= \mathbf{P}^{\omega}\big(\omega(\mathbf{C}^{s_1})^{-1}\underline{\hat{\mathbf{x}}}^{s_1} + (1-\omega)(\mathbf{C}^{s_2})^{-1}\underline{\hat{\mathbf{x}}}^{s_2}\big) , &(3.20)\\ \mathbf{P}^{\omega} &= \big(\omega(\mathbf{C}^{s_1})^{-1} + (1-\omega)(\mathbf{C}^{s_2})^{-1}\big)^{-1} , &(3.21)\end{aligned}$$

where the scalar parameter $\omega \in [0, 1]$ is to be optimized according to a criterion $\mathcal{J}\{\cdot\}$ with

$$\hat{\omega} = \arg\min_{\omega} \mathcal{J}\{\mathbf{P}^{\omega}\} . \tag{3.22}$$

Appropriate candidates for the criterion $\mathcal{J}\{\cdot\}$ are the trace or determinant [118]. The optimization of ω is necessary as unlike for the KF, CI does not provide a covariance $\mathbf{P}^{\omega}$ that is smaller than any other covariance bound in the positive definite sense. However, for any choice ω, the covariance provided by CI constitutes a bound [85] as illustrated by means of covariance ellipsoids in Figure 3.9. For the ease of notation, $\underline{\hat{\mathbf{x}}}^{\omega}$ and $\mathbf{P}^{\omega}$ are denoted as $\underline{\hat{\mathbf{x}}}$ and $\mathbf{P}$ in the sequel.

Readers familiar with linear fusion algorithms may have noticed the resemblance of (3.21) to the fusion of uncorrelated estimates described in Corollary 2.6. In fact, the CI formulas correspond to the

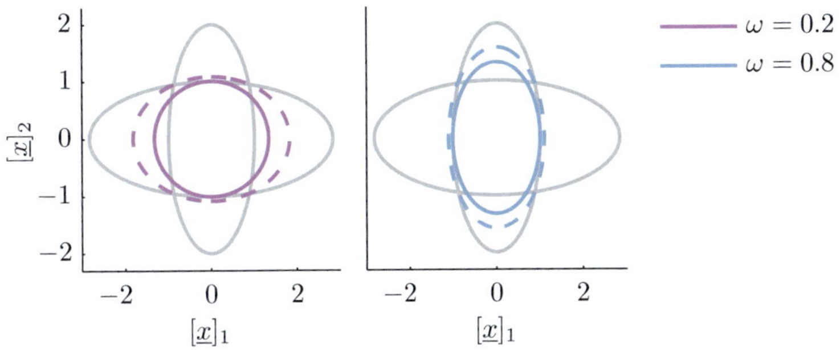

Figure 3.9: Bounds (dashed) provided by CI for prior covariances $\mathbf{C}^{s_1} = \mathrm{diag}(4, 1)$, $\mathbf{C}^{s_2} = \mathrm{diag}(1, 8)$, and $\omega \in \{0.2, 0.8\}$. The true covariances (solid) for $\mathbf{C}^{s_1 s_2} = \mathbf{I}$ are depicted as well.

convex combination of Corollary 2.6 for inflated local covariances $\frac{1}{\omega}\mathbf{C}^{s_1}$ and $\frac{1}{1-\omega}\mathbf{C}^{s_2}$. This similarity does not exist by chance. As it has been shown in [31], the inflation of local covariances yields bounds on the fused estimate for arbitrary unbiased fusion gains. Going one step further, joint covariance matrices delineated in Section 2.3.4 can be considered that permit the representation of local covariances and linear dependencies of arbitrary many estimates in a unified matrix. In particular, bounds on joint covariance matrices ensure consistency in the fusion according to the next statement.

Lemma 3.9 *Let $\bar{\mathbf{C}}$ denote the joint covariance matrix of estimates $\hat{\underline{\mathbf{x}}}^s$, $s \in \mathcal{S}$ and let $\bar{\mathbf{P}} \geq \bar{\mathbf{C}}$ denote a joint covariance matrix bound. For $\mathbf{F} = \left(\mathbf{F}^{s_1}, \ldots, \mathbf{F}^{s_S}\right)$ and $\sum_{s \in \mathcal{S}} \mathbf{F}^s = \mathbf{I}$, $\mathbf{P} = \mathbf{F}\bar{\mathbf{P}}(\mathbf{F})^\top$ constitutes a bound on the true covariance $\mathrm{E}\{(\sum_{s \in \mathcal{S}} \mathbf{F}^s \hat{\underline{\mathbf{x}}}^s - \underline{\mathbf{x}})^2\} = \mathbf{F}\bar{\mathbf{C}}(\mathbf{F})^\top$, i.e., $\mathbf{P} \geq \mathbf{F}\bar{\mathbf{C}}(\mathbf{F})^\top$.*

Proof. According to (2.27), the true covariance of the fused estimate is given by $\mathbf{F}\bar{\mathbf{C}}(\mathbf{F})^\top$. As $\mathbf{F} \in \mathbb{R}^{n_x \times |\mathcal{S}| \cdot n_x}$ and $n_x \leq |\mathcal{S}| \cdot n_x$, the claim follows immediately from Observation 7.7.2 in [76]. □

With Lemma 3.9, CI can be generalized to account for partially known correlations. Consider for example two estimates with errors $\underline{\mathbf{e}}^s = \hat{\underline{\mathbf{x}}}^s - \underline{\mathbf{x}}$, $s \in \{s_1, s_2\}$. When the errors $\underline{\mathbf{e}}^s$ are the **sums of two parts** $\underline{\mathbf{e}}^s = \underline{\mathbf{e}}_a^s + \underline{\mathbf{e}}_b^s$, and $\underline{\mathbf{e}}_a^{s_1}$ is independent from $\underline{\mathbf{e}}_b^{s_2}, \underline{\mathbf{e}}_a^{s_2}, \underline{\mathbf{e}}_b^{s_2}$ (analogously $\underline{\mathbf{e}}_a^{s_2}$), the joint covariance matrix is the sum of two matrices and unknown correlations must be bounded for one part only.

Interestingly, the additive error representation evolves naturally in linear models as noise terms can be considered independently of each other. This has been exploited in split CI [87,88] by proposing fusion formulas for the error decomposition into $\underline{\mathbf{e}}_a^s$ and $\underline{\mathbf{e}}_b^s$. An extension to multiple estimates and the explicit consideration of pedigree information has been proposed in [4,5].

When the errors $\underline{\mathbf{e}}_a^{s_1}$ and $\underline{\mathbf{e}}_a^{s_2}$ are not independent but correlated to a known extent, bounding techniques for two estimates have been presented in [68,120].

Alternatively, additional assumptions about the cross-covariance matrices can be introduced. In [67], a **scalar coefficient** has been utilized to confine the set of possible cross-covariance matrices and to provide smaller covariance bounds. This idea has been rediscovered and extended to include known correlations in [139]. An explicit consideration of estimation vectors $\hat{\underline{\mathbf{x}}}^{s_1}$ and $\hat{\underline{\mathbf{x}}}^{s_2}$ in the fusion process based on CI has been proposed by means of covariance union [163]. The motivation is to provide consistent estimates even if one of the prior quantities is spurious.

Further research has been carried out to reduce computational effort that is due to the optimization in (3.22). One path is to employ an optimization function that can be solved efficiently. In this context, approximations for trace and determinant optimizations have been proposed [48,118] and set theoretic measures have been utilized [167]. In [181], efficient techniques to solve the optimization with the trace or determinant as criterion are discussed and closed-form solutions for low-dimensional systems are provided.

3.2.1 - a Optimal Fusion under Unknown Correlations

For known joint covariance matrices, the LMMSE fusion for two estimates is uniquely specified by Theorem 2.5. In the following, an analogous result for the fusion under unknown correlations is obtained. To this end, CI is shown to be the solution of (3.19).

The optimality of CI has already been considered in literature. In [31] it has been shown that CI provides the smallest bound in terms of the trace when the fused covariance satisfies

$$\operatorname{tr}\{\mathbf{C}\} \stackrel{!}{=} \sqrt{\operatorname{tr}\{\mathbf{F}^{s_1}\mathbf{C}^{s_1}(\mathbf{F}^{s_1})^\top\}} + \sqrt{\operatorname{tr}\{\mathbf{F}^{s_2}\mathbf{C}^{s_2}(\mathbf{F}^{s_2})^\top\}} \ .$$

However, the assumed equation holds only for covariances that are inflated with a scalar factor. Thus, optimality of CI according (3.19) is not implied. In [163], tightness of CI in the joint space could be established. However, a connection between a tight bounding in the joint space and the optimality of the fused bound has not been provided. The following proof is conducted along the lines of [179]. It utilizes a result on the tight bounding of ellipsoidal intersections from [90,91].

Theorem 3.10 *Let $\mathcal{J}\{\cdot\}$ denote a strictly monotonically increasing cost function*[3]*. The linear combination of two priors that minimizes the covariance bound on the fused estimate under completely unknown correlations is specified by fusion gains*

$$\mathbf{F}^{s_1} = \hat{\omega}\mathbf{C}(\mathbf{C}^{s_1})^{-1} \text{ and } \mathbf{F}^{s_2} = (1-\hat{\omega})\mathbf{C}(\mathbf{C}^{s_2})^{-1} \tag{3.23}$$

with

$$\mathbf{C} = \left(\hat{\omega}(\mathbf{C}^{s_1})^{-1} + (1-\hat{\omega})(\mathbf{C}^{s_2})^{-1}\right)^{-1} , \tag{3.24}$$

where $\hat{\omega} = \arg\min_\omega \mathcal{J}\{\mathbf{C}\}$.

[3] A matrix function is called strictly monotonically increasing when it satisfies $\mathbf{C} > \tilde{\mathbf{C}} \Rightarrow \mathcal{J}\{\mathbf{C}\} > \mathcal{J}\{\tilde{\mathbf{C}}\}$.

PROOF. First, note that for each cross-covariance matrix $\mathbf{C}^{s_1 s_2}$, the covariance $\tilde{\mathbf{C}}$ of the LMMSE combination is obtained with Theorem 2.5. As optimality holds in the positive definite sense, the combination of estimates with other gains yields covariances $\mathbf{C}$ that are larger in the positive definite sense. In other words, the ellipsoid $\tilde{\mathcal{E}}$ of $\tilde{\mathbf{C}}$ is contained in the ellipsoids $\mathcal{E}$ of $\mathbf{C}$, i.e., $\tilde{\mathcal{E}} \subseteq \mathcal{E}$.

Hence, a necessary (but not sufficient!) condition for a covariance bound $\mathbf{P}$ is that it must be larger than $\tilde{\mathbf{C}}$ for all possible cross-covariance matrices in order to guarantee that $\mathbf{P} \geq \mathbf{C} \geq \tilde{\mathbf{C}}$, where $\mathbf{C}$ is the covariance of the fused estimate subject to the gains used in (3.19). According to statements (1) and (2) from [31], the set of optimal covariances for all possible cross-covariance matrices is described by the ellipsoidal intersection $\mathcal{E}^{s_1} \cap \mathcal{E}^{s_2}$, where $\mathcal{E}^{s}$ is the ellipsoid of the covariance $\mathbf{C}^{s}$, $s \in \{s_1, s_2\}$. From $\tilde{\mathbf{C}} \leq \mathbf{P}$, it follows that the ellipsoid that depicts the optimal bound $\tilde{\mathbf{P}}$ must contain the intersection $\mathcal{E}^{s_1} \cap \mathcal{E}^{s_2}$. Now, it has been proven by Kahan in [90, 91] that for all covariances $\mathbf{P}$ that are not represented by

$$\left(\omega(\mathbf{C}^{s_1})^{-1} + (1-\omega)(\mathbf{C}^{s_2})^{-1}\right)^{-1}, \omega \in [0,1],$$

a covariance $\tilde{\mathbf{P}}$ with $\tilde{\mathbf{P}} < \mathbf{P}$ whose ellipsoid encloses the intersection $\mathcal{E}^{s_1} \cap \mathcal{E}^{s_2}$ can be derived. It is a consequence of the strict monotonicity of the cost function that

$$\mathcal{J}\left\{\tilde{\mathbf{P}}\right\} < \mathcal{J}\left\{\mathbf{P}\right\}.$$

Hence, if gains can be found such that the true covariances are bounded by $\tilde{\mathbf{P}}$, the solution to (3.19) is found along these gains and bounds.

Indeed, it has been proven in [85] that for fixed $\omega \in (0,1)$, the gains (3.23) yield covariances that are conservatively bounded by $\tilde{\mathbf{P}}$. Therefore, the optimal gains and bounds are given by (3.23) and (3.24). For $\omega \in \{0,1\}$, one of the estimates is dismissed and so, a conservative bound is yielded trivially. Finally, with the optimal $\hat{\omega}$ the cost function is minimized, which concludes the proof. □

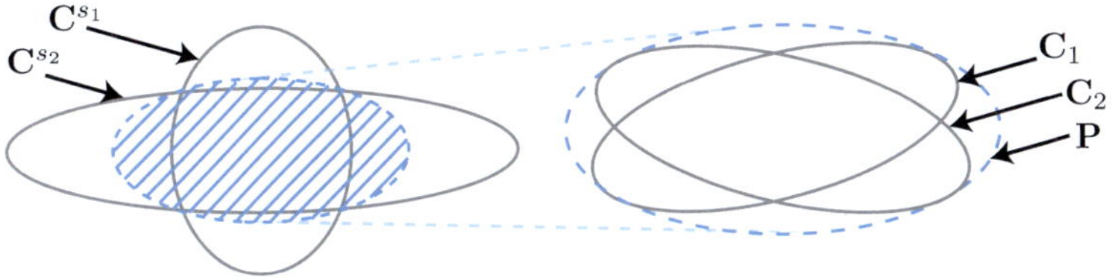

Figure 3.10: Emerging covariances and bounds in the fusion of two estimates. The prior covariances are depicted in grey and the CI bound in dashed blue. For specific cross-covariance matrices, the true (unknown) covariance for the fused estimate is illustrated in the right figure for two examples.

Obviously, gains and fused covariance from Theorem 3.10 correspond to the CI formulas (3.21). Now, let $\mathbf{C}$ and $\tilde{\mathbf{C}}$ denote two covariances with $\tilde{\mathbf{C}} < \mathbf{C}$. It follows from the positive definiteness of the matrix difference that the diagonal entries in $\mathbf{C}$ are larger than the ones of $\tilde{\mathbf{C}}$. Therefore, it holds

$$\operatorname{tr}\left\{\tilde{\mathbf{C}}\right\} < \operatorname{tr}\left\{\mathbf{C}\right\} ,$$

i.e., the trace is a strictly monotonically increasing cost function $\mathcal{J}\left\{\cdot\right\}$ and satisfies the assumption of Theorem 3.10. As the trace of the covariance denotes the MSE of the estimate, the following result is obtained.

Corollary 3.11 *The fusion of two estimates under completely unknown correlations that minimizes the MSE of the bound on the fused estimate is specified by CI.*

As demonstrated in the proof of Theorem 3.10, the covariance ellipsoid provided by CI encloses the covariance ellipsoids obtained by Theorem 2.5 for all possible cross-covariance matrices. Note, however, that for each cross-covariance matrix, the true covariance of the fused estimate obtained with CI gains (3.23) is larger than the counterpart obtained with LMMSE gains under known correlations, i.e., with Theorem 2.5. In particular, the true covariances obtained by CI are in general larger than the intersection $\mathcal{E}^{s_1} \cap \mathcal{E}^{s_2}$, as illustrated in Figure 3.10.

The extension of Theorem 3.10 to more than two estimates is not trivial. As shown in [91], ellipsoids provided by the natural extension of CI (c.f., Section 3.2.2) do not necessarily circumscribe the intersection of the prior covariance ellipsoids tightly. Therefore, it is possible that smaller bounds than the ones provided by CI can be derived for more than two estimates. A more detailed discussion on this topic is given in [179].

3.2.2 Generalized Covariance Intersection

In sensor network estimation, all information received at a particular time step is to be combined. In general, the number of packets exceeds two, which necessitates repeated applications of CI.

However, the bounds obtained with CI are optimal with respect to the considered cost function only and not in the positive definite sense. Hence, even if the same optimization criterion is utilized in each iteration, the bound is in general suboptimal after repeated applications of CI [174] as illustrated in Figure 3.11.

In the following, batch formulas are presented that enable the optimization of bounds for an arbitrary number of estimates and that outperform the sequential application of CI. Then, the inclusion of partially known cross-covariance matrices is facilitated based on a decomposition of error terms.

For CI, a generalization to more than two estimates has already been discussed in literature [31, 118, 163]. Let $\omega^{s_1}, \ldots, \omega^{s_S} \in [0, 1]$ denote

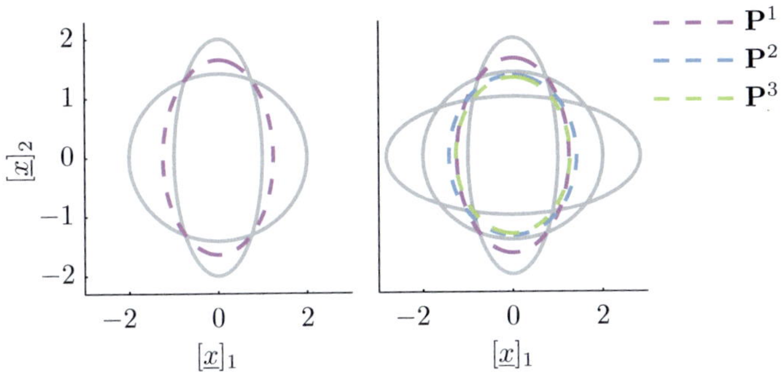

Figure 3.11: Bounds for three covariances $\mathbf{C}^{s_1} = \text{diag}\,(4,1)$, $\mathbf{C}^{s_2} = \text{diag}\,(2,4)$, and $\mathbf{C}^{s_3} = \text{diag}\,(1,8)$. In the left figure, the trace optimal bound $\mathbf{P}^1$ for $\mathbf{C}^{s_1}$ and $\mathbf{C}^{s_2}$ is depicted. When this bound is used as input for the fusion with $\mathbf{C}^{s_3}$, the bound $\mathbf{P}^2$ is obtained. However, a direct optimization over all covariances excludes information from $\mathbf{C}^{s_2}$ and yields $\mathbf{P}^3$ with $\text{tr}\left\{\mathbf{P}^3\right\} < \text{tr}\left\{\mathbf{P}^2\right\}$.

scalar weighting factors with $\sum_{s\in\mathcal{S}} \omega^s = 1$. The CI formulas generalize naturally to

$$\hat{\underline{\mathbf{x}}} = \mathbf{P} \sum_{s\in\mathcal{S}} \omega^s (\mathbf{C}^s)^{-1} \hat{\underline{\mathbf{x}}}^s \ , \tag{3.25}$$

with fused covariance bound

$$\mathbf{P} = \Big(\sum_{s\in\mathcal{S}} \omega^s (\mathbf{C}^s)^{-1} \Big)^{-1} \ , \tag{3.26}$$

where $\{\omega^s\}_{s\in\mathcal{S}}$ are obtained subject to the optimization criterion $\mathcal{J}\{\cdot\}$ according to

$$\{\hat{\omega}^s\}_{s\in\mathcal{S}} = \underset{\omega^{s_1},\ldots,\omega^{s_S}}{\arg\min}\ \mathcal{J}\{\mathbf{P}\} = \underset{\omega^{s_1},\ldots,\omega^{s_S}}{\arg\min}\ \mathcal{J}\left\{ \Big(\sum_{s\in\mathcal{S}} \omega^s (\mathbf{C}^s)^{-1} \Big)^{-1} \right\} . \tag{3.27}$$

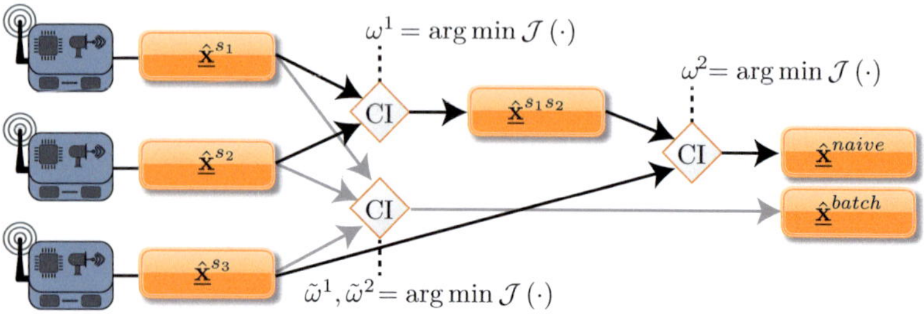

Figure 3.12: The naïve sequential application of CI versus the batch processing that optimizes all $\{\omega^s\}_{s=s_1,s_2}$ at once.

It is easy to verify that the formulas boil down to CI when $|\mathcal{S}| = 2$ and provide different weights than a sequential optimization when more than two estimates are combined as illustrated in Figure 3.12.

It has already been mentioned for two estimates that a bound on the joint covariance matrix is obtained by means of inflated block diagonal covariances. The corresponding statement for arbitrary many estimates is given as follows.

Lemma 3.12 *Let $\bar{\mathbf{C}}$ denote a joint covariance matrix with covariances $\mathbf{C}^s$, $s \in \mathcal{S}$ on the block diagonal. The matrix $\bar{\mathbf{P}}$ defined as*

$$\bar{\mathbf{P}} = \begin{pmatrix} \frac{1}{\omega^{s_1}}\mathbf{C}^{s_1} & \cdots & \mathbf{0} \\ \vdots & \ddots & \vdots \\ \mathbf{0} & \cdots & \frac{1}{\omega^{s_S}}\mathbf{C}^{s_S} \end{pmatrix} \tag{3.28}$$

with $\omega^s > 0$, $\sum_{s\in\mathcal{S}} \omega^s = 1$ is a bound on $\bar{\mathbf{C}}$.

PROOF. The proof follows [178]. Let $\bar{\mathbf{C}}^{\tilde{\mathcal{S}}}$ and $\bar{\mathbf{P}}^{\tilde{\mathcal{S}}}$ denote the joint covariance matrix and the corresponding bound of the first $|\tilde{\mathcal{S}}| \leq |\mathcal{S}|$

entries, i.e.,

$$\bar{\mathbf{C}}^{\tilde{\mathcal{S}}} = \begin{pmatrix} \mathbf{C}^{s_1} & \dots & \mathbf{C}^{s_1 s_{\tilde{\mathcal{S}}}} \\ \vdots & \ddots & \vdots \\ \mathbf{C}^{s_{\tilde{\mathcal{S}}} s_1} & \dots & \mathbf{C}^{s_{\tilde{\mathcal{S}}}} \end{pmatrix}$$

and

$$\bar{\mathbf{P}}^{\tilde{\mathcal{S}}} = \begin{pmatrix} \frac{1}{\omega^{s_1}} \mathbf{C}^{s_1} & \dots & \mathbf{0} \\ \vdots & \ddots & \vdots \\ \mathbf{0} & \dots & \frac{1}{\omega^{s_{\tilde{\mathcal{S}}}}} \mathbf{C}^{s_{\tilde{\mathcal{S}}}} \end{pmatrix} .$$

It is proven by induction that $\bar{\omega}^{\tilde{\mathcal{S}}} \bar{\mathbf{P}}^{\tilde{\mathcal{S}}} \geq \bar{\mathbf{C}}^{\tilde{\mathcal{S}}}$ with $\bar{\omega}^{\tilde{\mathcal{S}}} = \sum_{s \in \tilde{\mathcal{S}}} \omega^s$. Then, the claim follows immediately with $\bar{\omega}^{\mathcal{S}} = 1$.

For $|\tilde{\mathcal{S}}| = 1$, it holds $\bar{\omega}^1 \bar{\mathbf{P}}^1 = \frac{\omega^{s_1}}{\omega^{s_1}} \mathbf{C}^{s_1} \geq \mathbf{C}^{s_1}$. For the induction step, let the claim be satisfied for $|\tilde{\mathcal{S}}| - 1$. Then,

$$\begin{aligned} \bar{\omega}^{\tilde{\mathcal{S}}} \bar{\mathbf{P}}^{\tilde{\mathcal{S}}} - \bar{\mathbf{C}}^{\tilde{\mathcal{S}}} &= \bar{\omega}^{\tilde{\mathcal{S}}} \begin{pmatrix} \frac{1}{\bar{\omega}^{\tilde{\mathcal{S}}-1}} \bar{\omega}^{\tilde{\mathcal{S}}-1} \bar{\mathbf{P}}^{\tilde{\mathcal{S}}-1} & \mathbf{0} \\ \mathbf{0} & \frac{1}{\omega^{\tilde{\mathcal{S}}}} \mathbf{C}^{s_{\tilde{\mathcal{S}}}} \end{pmatrix} - \bar{\mathbf{C}} \\ &= \begin{pmatrix} \frac{\bar{\omega}^{\tilde{\mathcal{S}}}}{\bar{\omega}^{\tilde{\mathcal{S}}-1}} \bar{\omega}^{\tilde{\mathcal{S}}-1} \bar{\mathbf{P}}^{\tilde{\mathcal{S}}-1} - \bar{\mathbf{C}}^{\tilde{\mathcal{S}}-1} & -\bar{\mathbf{C}}^{(\tilde{\mathcal{S}}-1)\tilde{\mathcal{S}}} \\ -\bar{\mathbf{C}}^{\tilde{\mathcal{S}}(\tilde{\mathcal{S}}-1)} & \frac{\bar{\omega}^{\tilde{\mathcal{S}}}}{\omega^{\tilde{\mathcal{S}}}} \mathbf{C}^{s_{\tilde{\mathcal{S}}}} - \mathbf{C}^{s_{\tilde{\mathcal{S}}}} \end{pmatrix} , \end{aligned}$$

where $\bar{\mathbf{C}}^{(\tilde{\mathcal{S}}-1)\tilde{\mathcal{S}}}$ denotes an unknown cross-covariance matrix. According to the induction hypothesis, it holds $\bar{\omega}^{\tilde{\mathcal{S}}-1} \bar{\mathbf{P}}^{\tilde{\mathcal{S}}-1} \geq \bar{\mathbf{C}}^{\tilde{\mathcal{S}}-1}$. Thus, it follows with $\frac{\bar{\omega}^{\tilde{\mathcal{S}}}}{\bar{\omega}^{\tilde{\mathcal{S}}-1}} \geq 0$,

$$\bar{\omega}^{\tilde{\mathcal{S}}} \bar{\mathbf{P}}^{\tilde{\mathcal{S}}} - \bar{\mathbf{C}}^{\tilde{\mathcal{S}}} \geq \begin{pmatrix} \frac{\bar{\omega}^{\tilde{\mathcal{S}}} - \bar{\omega}^{\tilde{\mathcal{S}}-1}}{\bar{\omega}^{\tilde{\mathcal{S}}-1}} \bar{\mathbf{C}}^{\tilde{\mathcal{S}}-1} & -\bar{\mathbf{C}}^{(\tilde{\mathcal{S}}-1)\tilde{\mathcal{S}}} \\ -\bar{\mathbf{C}}^{\tilde{\mathcal{S}}(\tilde{\mathcal{S}}-1)} & \frac{\bar{\omega}^{\tilde{\mathcal{S}}} - \omega^{\tilde{\mathcal{S}}}}{\omega^{\tilde{\mathcal{S}}}} \mathbf{C}^{s_{\tilde{\mathcal{S}}}} \end{pmatrix} .$$

According to Theorem 7.7.6 from [76], the difference is positive definite, if and only if

$$\bar{\mathbf{C}}^{\tilde{\mathcal{S}}(\tilde{\mathcal{S}}-1)} \frac{\bar{\omega}^{\tilde{\mathcal{S}}-1}}{\bar{\omega}^{\tilde{\mathcal{S}}} - \bar{\omega}^{\tilde{\mathcal{S}}-1}} (\bar{\mathbf{C}}^{\tilde{\mathcal{S}}-1})^{-1} \bar{\mathbf{C}}^{(\tilde{\mathcal{S}}-1)\tilde{\mathcal{S}}} < \frac{\bar{\omega}^{\tilde{\mathcal{S}}} - \omega^{\tilde{\mathcal{S}}}}{\omega^{\tilde{\mathcal{S}}}} \mathbf{C}^{s_{\tilde{\mathcal{S}}}} .$$

With $\bar{\omega}^{\tilde{S}} = \bar{\omega}^{\tilde{S}-1} + \omega^{\tilde{S}}$, the scalar factor simplifies to

$$\frac{\bar{\omega}^{\tilde{S}-1}}{\bar{\omega}^{\tilde{S}} - \bar{\omega}^{\tilde{S}-1}} = \frac{\bar{\omega}^{\tilde{S}-1}}{\omega^{\tilde{S}}} = \frac{\bar{\omega}^{\tilde{S}} - \omega^{\tilde{S}}}{\omega^{\tilde{S}}} .$$

Therefore, it holds $\bar{\omega}^{\tilde{S}}\bar{\mathbf{P}}^{\tilde{S}} \geq \bar{\mathbf{C}}^{\tilde{S}}$ for positive semi-definite $\mathbf{C}^{s\tilde{s}}$. □

Note that this result is more general than the consistency statements in [31, 85, 118, 146], as consistency of the fused estimate follows from consistency in the joint space according to Lemma 3.9. Applied to the batch CI formulas, the next corollary is obtained.

Corollary 3.13 *Let* $\hat{\underline{\mathbf{x}}}$ *and* $\mathbf{P}$ *denote the CI fusion result from* (3.25) *and* (3.26). *When* $\mathbf{C}^s = \mathrm{E}\{(\hat{\underline{\mathbf{x}}}^s - \underline{\mathbf{x}})^2\}$, $s \in \mathcal{S}$, *then* $\mathbf{P} \geq \mathrm{E}\{(\hat{\underline{\mathbf{x}}} - \underline{\mathbf{x}})^2\}$.

PROOF. Let $\bar{\mathbf{C}}$ denote the true joint covariance matrix with $\mathbf{C}^s$ on the block diagonal and $\bar{\mathbf{P}}$ the bound from Lemma 3.12. Consider the fusion matrix $\mathbf{F} = \begin{pmatrix} \omega^{s_1}\mathbf{P}(\mathbf{C}^{s_1})^{-1} & \dots & \omega^{s_S}\mathbf{P}(\mathbf{C}^{s_S})^{-1} \end{pmatrix}$. Then, the claim follows immediately with Lemma 3.9. □

3.2.2 - a Partially known Cross-covariance Matrices

Now, let the errors of the prior estimates be the sums of two uncorrelated parts each, i.e.,

$$\underline{\mathbf{e}}^s = \hat{\underline{\mathbf{x}}}^s - \underline{\mathbf{x}} = \underline{\mathbf{e}}_a^s + \underline{\mathbf{e}}_b^s \text{ with } \mathrm{E}\{\underline{\mathbf{e}}_a^s(\underline{\mathbf{e}}_b^{\tilde{s}})^\top\} = \mathbf{0} ,$$

and let $\bar{\underline{\mathbf{e}}} = \begin{pmatrix} (\underline{\mathbf{e}}^{s_1})^\top & \cdots & (\underline{\mathbf{e}}^{s_S})^\top \end{pmatrix}^\top$ denote the joint space error with decomposition $\bar{\underline{\mathbf{e}}} = \bar{\underline{\mathbf{e}}}_a + \bar{\underline{\mathbf{e}}}_b$. As error parts $\underline{\mathbf{e}}_a$ and $\underline{\mathbf{e}}_b$ are independent from each other, the joint covariance matrix is given by

$$\bar{\mathbf{C}} = \mathrm{E}\{\bar{\underline{\mathbf{e}}}(\bar{\underline{\mathbf{e}}})^\top\} = \mathrm{E}\{\bar{\underline{\mathbf{e}}}_a(\bar{\underline{\mathbf{e}}}_a)^\top\} + \mathrm{E}\{\bar{\underline{\mathbf{e}}}_b(\bar{\underline{\mathbf{e}}}_b)^\top\} = \bar{\mathbf{C}}_a + \bar{\mathbf{C}}_b ,$$

where $\bar{\mathbf{C}}_a$ and $\bar{\mathbf{C}}_b$ are the joint covariance matrices that account for the error terms $\underline{\mathbf{e}}_a$ and $\underline{\mathbf{e}}_b$ respectively.

Starting from the premise that all (cross-)covariance matrices between $\underline{\mathbf{e}}_a^s$ and $\underline{\mathbf{e}}_a^{\tilde{s}}$ as well as the (local) errors $\mathrm{E}\{\underline{\mathbf{e}}_b^s(\underline{\mathbf{e}}_b^s)^\top\}$ are known, the joint covariance matrix $\bar{\mathbf{C}}_a$ and the block diagonal matrices of $\bar{\mathbf{C}}_b$ can be calculated. Let $\bar{\mathbf{P}}_b$ denote the joint covariance matrix bound of $\bar{\mathbf{C}}_b$ from Lemma 3.12. Then, a **bound for the (overall) joint covariance matrix** is given by

$$\bar{\mathbf{P}} := \bar{\mathbf{C}}_a + \bar{\mathbf{P}}_b \geq \bar{\mathbf{C}}_a + \bar{\mathbf{C}}_b = \bar{\mathbf{C}} \ , \tag{3.29}$$

which, in turn, specifies a bound on the fused covariance for any linear combination of estimates by means of Lemma 3.9. The optimal fusion gains for the additive covariance representation are specified in the next statement.

Theorem 3.14 *Let estimates $\hat{\underline{\mathbf{x}}}^s$, $s \in \mathcal{S}$ with joint covariance matrix decomposition $\bar{\mathbf{C}} = \bar{\mathbf{C}}_a + \bar{\mathbf{C}}_b$ be given. Then, $\bar{\mathbf{P}} = \bar{\mathbf{C}}_a + \bar{\mathbf{P}}_b$ with $\bar{\mathbf{P}}_b$ from Lemma 3.12 provides a joint covariance matrix bound and the fusion gains $\mathbf{F} = \begin{pmatrix}\mathbf{F}^{s_1} & \dots & \mathbf{F}^{s_S}\end{pmatrix}$ that minimize the bound $\mathbf{P}$ on the fused estimate are obtained by*

$$\mathbf{F} = \begin{pmatrix}\mathbf{F}^{s_1} & \dots & \mathbf{F}^{s_S}\end{pmatrix} = \mathbf{P}\underline{\mathbf{1}}(\bar{\mathbf{P}})^+ \ \textit{with}\ \mathbf{P} = ((\underline{\mathbf{1}})^\top(\bar{\mathbf{P}})^+\underline{\mathbf{1}})^+ \ , \tag{3.30}$$

where $\bar{\mathbf{P}}_b$ is chosen subject to the optimization problem

$$\min_{\omega^{s_1},\dots,\omega^{s_S}} \mathcal{J}\{\mathbf{P}\} \ , \ \omega^s > 0 \ , \ \sum_{s\in\mathcal{S}} \omega^s = 1 \ . \tag{3.31}$$

Proof. The proof follows [178]. The bound provided by Theorem 3.14 is the result of a fusion operation with joint covariance matrix $\bar{\mathbf{C}} = \bar{\mathbf{C}}_a + \bar{\mathbf{P}}_b$. For a given joint covariance matrix with fixed weights, the gains that provide the optimal covariance in the positive definite order are given by Theorem 2.5. An application of the formula leads to the fused covariance bound $\mathbf{P} = ((\underline{\mathbf{1}})^\top(\bar{\mathbf{C}}_a + \bar{\mathbf{P}}_b)^+\underline{\mathbf{1}})^+$, and the specified fusion gains. The given optimization problem for $\{\omega^s\}_{s\in\mathcal{S}}$ calculates the optimal weights subject to the cost function $\mathcal{J}\{\cdot\}$. □

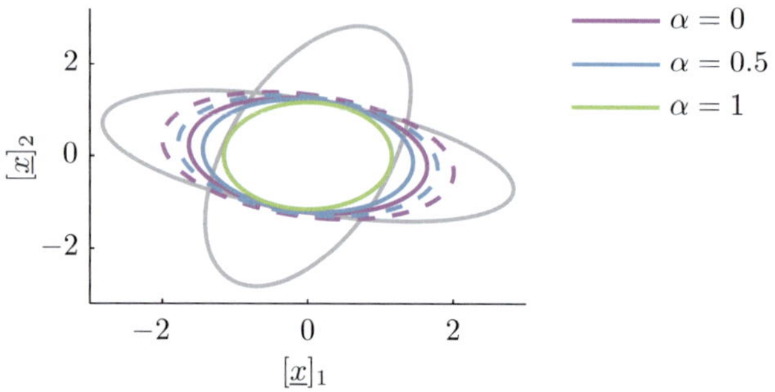

Figure 3.13: The ellipsoids for the covariances $\mathbf{C}^{s_1}$ and $\mathbf{C}^{s_2}$ as well as for the covariance of the fused estimate $\mathbf{C}$ subject to different degrees of knowledge about cross-covariance matrices.

Theorem 3.14 constitutes the optimal fusion method for the theory in [68, 120] and unifies the LMMSE fusion under known correlations from Theorem 2.5 with bounding techniques under unknown correlations. In particular, some special cases are covered:

- For $\bar{\mathbf{C}}_a = \mathbf{0}$, the joint covariance matrix bound is given by $\bar{\mathbf{P}} = \bar{\mathbf{P}}_b$, i.e., batch CI from (3.25) and (3.26) is obtained.
- For $\bar{\mathbf{C}}_a$ block diagonal, the known error parts $\underline{\bar{\mathbf{e}}}_a$ are uncorrelated, and the optimization corresponds to split CI [88].
- For $\bar{\mathbf{C}}_b = \mathbf{0}$, no bounding is necessary and the LMMSE fusion under known correlations from Theorem 2.5 is obtained.

Note that $\bar{\mathbf{C}}_a$ comprises known correlations and $\bar{\mathbf{C}}_b$ aggregates errors terms with unclear dependencies. Hence, a worst-case **assessment of the tightness** of the covariance $\bar{\mathbf{C}} = \bar{\mathbf{C}}_a + \bar{\mathbf{C}}_b$ is given by

$$\alpha = \frac{\operatorname{tr}\left\{\bar{\mathbf{C}}_a\right\}}{\operatorname{tr}\left\{\bar{\mathbf{C}}_a\right\} + \operatorname{tr}\left\{\bar{\mathbf{C}}_b\right\}}, \alpha \in [0, 1] . \tag{3.32}$$

For $\alpha = 1$, the covariance of the fused estimate is exact and for $\alpha = 0$, (conservative) CI bounds are provided. In general, the smaller α is, the more inexact is the covariance of the fused estimate provided in Theorem 3.14. This relation is illustrated in the next example.

Example 3.6: CI with Partially Known Cross-covariance Matrix

Consider two estimates defined by (cross-)covariance matrices

$$\mathbf{C}^{s_1} = \begin{pmatrix} 8 & 2 \\ 2 & 2 \end{pmatrix}, \mathbf{C}^{s_2} = \begin{pmatrix} 2 & -2 \\ -2 & 8 \end{pmatrix}, \text{ and } \mathbf{C}^{s_1 s_2} = \begin{pmatrix} 2 & 0 \\ 0 & 2 \end{pmatrix}.$$

The optimization of fusion gains according to Theorem 3.14 depends on the knowledge about the cross-covariance matrices.

In this example, the cases with no, partial, and full knowledge about cross-covariance matrices are examined. To this end, the same joint covariance matrix $\bar{\mathbf{C}}$ with the following decompositions into $\bar{\mathbf{C}}_a$ and $\bar{\mathbf{C}}_b$ is considered:

- No knowledge: $\alpha = 0$: $\bar{\mathbf{C}}_a = \mathbf{0}$, $\bar{\mathbf{C}}_b = \mathbf{C}$.
- Partial knowledge: $\alpha = 0.5$: $\bar{\mathbf{C}}_a = 0.5 \cdot \mathbf{C}$, $\bar{\mathbf{C}}_b = 0.5 \cdot \mathbf{C}$.
- Full knowledge: $\alpha = 1$: $\bar{\mathbf{C}}_a = \mathbf{C}$, $\bar{\mathbf{C}}_b = \mathbf{0}$.

In Figure 3.13, the bounds and true covariances of the fused estimate obtained with gains (3.30) and $\omega = 0.2$ are depicted. The more knowledge about cross-covariance matrices is available to the estimator, the smaller are the bounds.

In fact, for the considered scalar decomposition of $\bar{\mathbf{C}}$ into $\bar{\mathbf{C}}_a$ and $\bar{\mathbf{C}}_b$, the bound and true covariances shrink monotonically with increasing α. By examining other decompositions, a smooth transition from CI ($\alpha = 0$) to LMMSE fusion results ($\alpha = 1$) can be observed. ■

Apart from the special case that $\bar{\mathbf{C}}_b = \mathbf{0}$, it is necessary to optimize the parameters $\{\omega^s\}_{s\in\mathcal{S}}$ in (3.31). Indeed, as discussed in Section 3.2.1, closed-form equations for arbitrary dimensions and optimization criteria such as the trace or determinant are not even available for two estimates. Thus, for more than two estimates, a **multivariate optimization problem** has to be solved with numerical methods, which in general involves a considerable computational effort. Fortunately, as motivated in Figure 3.14 and proven in Appendix C, the optimization is convex for the trace as optimization criterion.

Theorem 3.15 *For $\mathcal{J}\{\cdot\} = \mathrm{tr}\{\cdot\}$, the optimization of $\{\omega^s\}_{s\in\mathcal{S}}$ in Theorem 3.14 is an equality constrained convex optimization problem.*

For convex optimization problems, local minima are globally optimal so that numerical optimization techniques, e.g., gradient descend, converge to the desired solution.

Moreover, powerful toolboxes are available that efficiently compute convex optimization problems [20, 61]. As it has been argued above, batch CI is a special case of Theorem 3.14 so that Theorem 3.15 applies.

Corollary 3.16 *The CI optimization from (3.27) with $\mathcal{J}\{\cdot\}=\mathrm{tr}\{\cdot\}$ is convex.*

As it has been demonstrated, CI bounding techniques can be used to generalize the LMMSE fusion theory in terms of the knowledge about correlations. More precisely, knowledge about cross-covariance matrices can be included and the optimal fusion gains in the considered framework can be obtained with Theorem 3.14. The necessary optimization is convex so that efficient implementations can be used.

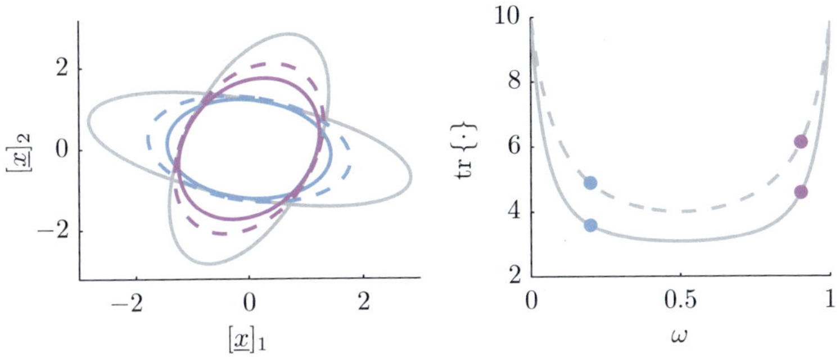

Figure 3.14: In the left figure, covariances and bounds of Example 3.6 with $\alpha = 0.5$ for $\mathbf{C}^{s_1}$, $\mathbf{C}^{s_2}$, and $\mathbf{C}$ with $\omega \in \{0.2, 0.9\}$ are depicted. In the right figure, the traces of $\mathbf{C}$ and $\mathbf{P}$ are plotted for all relevant ω.

3.2.3 Aggregation of Independent Noise Terms

In order to optimize the bound on the fused estimate, the part of the covariance that needs to be inflated, i.e., $\bar{\mathbf{C}}_b$ must be minimized. In the following, the conditional independence of locally observed measurements is exploited to derive a decomposition of the local covariance at each node into a part that represents exclusively local information and a remainder. As error terms comprised in the former are independent from all other noise terms, the corresponding joint covariance matrix is block diagonal and can be used to improve the bound in Theorem 3.14. In fact, the collection of independent terms has originally been proposed with split CI [88] such that the following considerations are an application of the split CI theory to linear estimation.

Before an aggregation of noise covariances is presented, the evolution of (cross-)covariance matrices in sensor networks is analyzed. For this purpose, the error decomposition from Section 3.1.1 is applied to the cross-covariance matrix definition $\mathbf{C}^{s_1 s_2} = \mathrm{E}\{\underline{\mathbf{e}}^{s_1}(\underline{\mathbf{e}}^{s_2})^\top\}$. The relation between occurring quantities is depicted in Figure 3.15.

At **initialization**, the cross-covariance matrix is determined by the correlation of the estimates. For independent estimates, the cross-covariance matrix is $\mathrm{E}\{\underline{\mathbf{e}}_0^{s_1}(\underline{\mathbf{e}}_0^{s_2})^\top\} = \mathbf{0}$. In the **prediction** step, the common process noise $\{\underline{\mathbf{w}}_k\}_{k\in\mathcal{K}}$ is added to the true system, leading to

$$\mathbf{C}_{k+1}^{s_1s_2} = \mathbf{A}_k\mathbf{C}_k^{s_1s_2}(\mathbf{A}_k)^\top + \mathbf{Q}_k \,. \tag{3.33}$$

In the **filtering** operation, estimates $\hat{\underline{\mathbf{x}}}^s$ are transformed with $\mathbf{L}^s$ and measurement noise with $\mathrm{E}\{(\hat{\underline{\mathbf{x}}}^s - \underline{\mathbf{x}})\underline{\mathbf{v}}^s\} = \mathbf{0}$ is added such that the cross-covariance matrix follows as

$$\mathbf{C}_{k|k}^{s_1s_2} = \mathbf{L}_k^{s_1}\mathbf{C}_k^{s_1s_2}(\mathbf{L}_k^{s_2})^\top \,. \tag{3.34}$$

The description of concurrent transformations at multiple nodes in the **fusion** operation amounts to a complicated formula that is the result of the matrix transformation in (2.36). If only the fusion of estimates at node s_1 is considered, this formula simplifies to

$$\mathbf{C}^{s_1s_2} = \sum_{s\in\mathcal{S}} \mathbf{F}^s\mathbf{C}^{ss_2} \,. \tag{3.35}$$

With the independence of individual noise terms, i.e., $\mathrm{E}\{\underline{\boldsymbol{\psi}}_i(\underline{\boldsymbol{\psi}}_{\tilde{i}})^\top\} = \mathbf{0}$ for $i \neq \tilde{i}$, the same considerations that led to the sum representation of the error (3.7) also permit specifying a sum representation of (cross-)covariance matrices.

Theorem 3.17 *The (cross-)covariance matrices of linear estimators in linear systems according to Definition 2.1 are given by*

$$\mathbf{C}^{s_1s_2} = \sum_{i\in\mathcal{I}^{s_1}\cap\mathcal{I}^{s_2}} \mathbf{T}_i^{s_1}\mathbf{C}_i^{\nu}(\mathbf{T}_i^{s_2})^\top \,, \tag{3.36}$$

with transformation matrices $\mathbf{T}_i^s$ *from* (3.7) *and noise covariances* $\mathbf{C}_i^{\nu} = \mathrm{E}\{\underline{\boldsymbol{\psi}}_i(\underline{\boldsymbol{\psi}}_i)^\top\}$, $\underline{\boldsymbol{\psi}}_i \in \{\underline{\mathbf{v}}_k^s, \underline{\mathbf{w}}_k, \underline{\mathbf{e}}_0^s \mid s \in \mathcal{S}\ , k \in \mathcal{K}\}$.

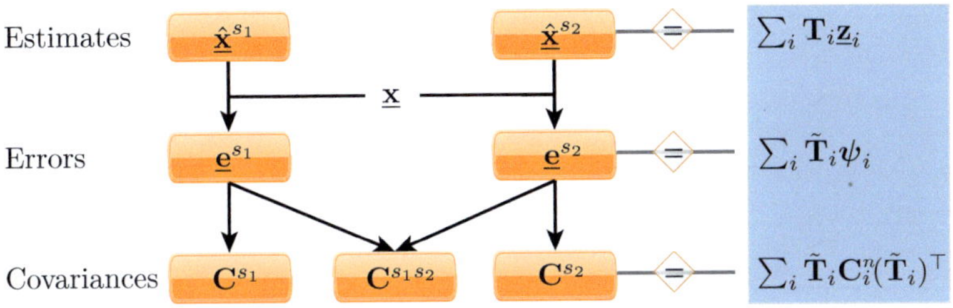

Figure 3.15: Relations between estimates, true state, error processes, and covariances. By means of estimates and the true state, error processes are determined, which, in turn, define (cross-)covariances.

Note that the transformation matrices $\mathbf{T}_i^s$ comprise all linear transformations that have been applied to the estimator since the respective noise term has emerged. In particular, it is possible to calculate the transformation matrices recursively at the sensors without global model knowledge, e.g., in the prediction step by a multiplication with $\mathbf{A}$.

In the subsequent considerations, it is exploited that measurement noise covariances do not emerge in cross-covariance matrices as long as estimates have not been fused. Hence, the idea is to **classify terms** in the sum decomposition from Theorem 3.17 into a part $\mathbf{C}_a^s$ that consists of linearly transformed measurement noise covariances and a part $\mathbf{C}_b^s$ that accounts for all process noise covariances. Then, both covariances can be processed recursively with standard covariance formulas (2.22), (2.24) and (2.27) at the sensors.

Now, consider the fusion of two estimates $\hat{\underline{\mathbf{x}}}^{s_1}$ and $\hat{\underline{\mathbf{x}}}^{s_2}$ with independently observed information. As measurement noise terms stored in $\mathbf{C}_a^s$ are independent from all other noise terms in the sensor network, the corresponding joint covariance matrix $\bar{\mathbf{C}}_a$ is given by the block diagonal matrix

$$\bar{\mathbf{C}}_a = \begin{pmatrix} \mathbf{C}_a^{s_1} & \mathbf{0} \\ \mathbf{0} & \mathbf{C}_a^{s_2} \end{pmatrix} .$$

The cross-covariance matrices between $\mathbf{C}_b^{s_1}$ and $\mathbf{C}_b^{s_2}$ depend on transformations of both sensors and are assumed unknown. Based on the joint covariance matrix bound $\bar{\mathbf{P}} = \bar{\mathbf{C}}_a + \bar{\mathbf{P}}_b$, where $\bar{\mathbf{P}}_b$ is obtained from $\mathbf{C}_b^s$ with Lemma 3.12, fusion gains can be derived with Theorem 3.14. Hence, by assembling noise terms in **two separate covariances**, independent information can be explicitly considered.

Indeed, after the fusion in time step k, the measurements from sensor s_1 are also comprised in the estimate at sensor s_2. Thus, the measurement noise terms $\underline{\mathbf{v}}_{\tilde{k}}^{s_1}$, $\tilde{k} \leq k$ at sensor s_1 are correlated with other errors in the sensor network. Hence, it is necessary to set $\mathbf{C}_b = \mathbf{P}$ and $\mathbf{C}_a = \mathbf{0}$ to ensure the block diagonality of the joint covariance matrix $\bar{\mathbf{C}}_a$. In subsequent time steps, measurement noise terms are added to $\mathbf{C}_a$ such that in the fusion of time step $k+1$, the independence of $\underline{\mathbf{v}}_{k+1}^{s_1}$ from other noise terms can be exploited. The processing is summarized in Figure 3.16.

For the considered scenario with two sensors and for distributed estimation systems, the loss of correlation information in the fusion is not critical as the estimates are highly correlated (if not even equivalent) after the fusion and for high correlations the true covariance is almost as large as the bounded covariance. Unfortunately, the effect is more striking in **decentralized estimation**. Consider three sensors s_1, s_2, and s_3 that observe the same system. Let sensors s_1 and s_2 communicate in time step k, and let sensors s_1 and s_3 communicate in time step $k+2$. Then, $\mathbf{C}_a^{s_1}$ is set to $\mathbf{0}$ in the fusion with sensor s_2 so that in the subsequent fusion with sensor s_3 almost all noise terms must be inflated. This is reasonable when sensor s_2 communicates to s_3 in time step $k+1$. However, otherwise, the measurement noises of sensors s_1 and s_3 might be independent so that the achieved bound is more conservative than necessary.

A generalization of split CI that permits the processing of several covariances to account for independent measurement noise at each sensor has been proposed in [4,5]. However, independence is difficult to achieve in sensor networks as often (weak) correlations emerge

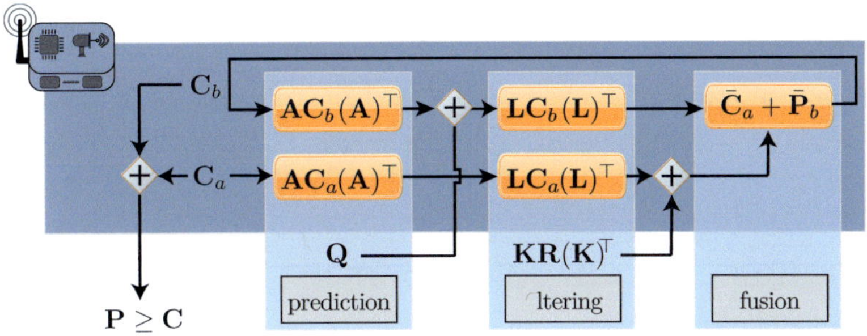

Figure 3.16: Overview over the processing of covariance bounds for the independent noise separation. In the fusion, estimates are exchanged and so, the measurement noise terms must be bounded subsequently in $\mathbf{C}_b$. The sum of $\mathbf{C}_a$ and $\mathbf{C}_b$ constitutes a covariance bound $\mathbf{P}$.

between estimates due to past data exchanges or the diffusion of information through different communication paths. Therefore, in decentralized estimation with an arbitrary communication, the proposed generalization achieves similar results as split CI and hence, brings only little advantage in terms of minimizing the bound.

3.2.4 Individual Noise Decomposition

Motivated by the limitations of split CI, a more general decomposition scheme that is applicable to sensor network estimation has been proposed in [178]. The idea is to process individual noise covariances at the sensors. As it turns out, such bookkeeping allows reconstructing cross-covariance matrices based on recursively processed matrices even if global model knowledge is not available. In particular, it is possible to retain the decomposition structure in the fusion. However, the processing comes with an infinitely growing computational and communication effort so that in a second step a bounding technique is proposed to confine the effort at the costs of optimality.

3.2.4 - a Exact Reconstruction of Cross-covariance Matrices

Considering the covariance decomposition into transformed noise covariances from Theorem 3.17, a technique to reconstruct cross-covariance matrices is to store square root terms $\sqrt{\mathbf{C}_i^s}$ that comprise transformations $\mathbf{T}_i^s$ applied to noise covariances $\mathbf{C}_i^\nu$ such that

$$\sqrt{\mathbf{C}_i^{s_1}}\sqrt{\mathbf{C}_i^{s_2}}^\top = \mathbf{T}_i^{s_1}\mathbf{C}_i^\nu(\mathbf{T}_i^{s_2})^\top \ .$$

Beginning with a unique square root of the noise covariance $\mathbf{C}_i^\nu$ such as the Cholesky decomposition, the transformation matrices, and, consequently, the square roots $\sqrt{\mathbf{C}_i^{s_1}}$, can be calculated recursively at the sensors without global model knowledge. According to (3.3), it is, in particular, feasible to combine noise covariances in the fusion operation according to

$$\sqrt{\mathbf{C}_i} = \sum_{s \in \mathcal{S}} \mathbf{F}^s \sqrt{\mathbf{C}_i^s} \ .$$

Hence, an algorithm for the exact reconstruction of cross-covariance matrices is to process a set $\mathcal{C}^s$ of tuples $(\sqrt{\mathbf{C}_i^s}, i)$ with indices $i \in \mathcal{I}$ at each sensor, exchange these sets along with the estimates, and reconstruct covariances by means of the sum representation from Theorem 3.17 according to

$$\mathbf{C}^s = ||\mathcal{C}^s|| := \sum_{(\sqrt{\mathbf{C}_i^s}, i) \in \mathcal{C}^s} \sqrt{\mathbf{C}_i^s}\sqrt{\mathbf{C}_i^s}^\top \ , \tag{3.37}$$

and cross-covariance matrices according to the corresponding sum over the common indices. The proposed processing is summarized in Figure 3.17, where $\mathbf{T}\mathcal{C}^s$ denotes the short form

$$\mathbf{T}\mathcal{C}^s := \{(\mathbf{T}\sqrt{\mathbf{C}_i^s}, i) \mid (\mathbf{C}_i^s, i) \in \mathcal{C}^s\} \ .$$

Let $\{n_\nu^i\}_{i \in \mathcal{I}}$ denote the dimensions of noise terms $\{\underline{\psi}_i\}_{i \in \mathcal{I}}$. The square root matrices have dimension $n_x \times n_\nu^i$ such that for low-dimensional measurements the number of stored entries per term is

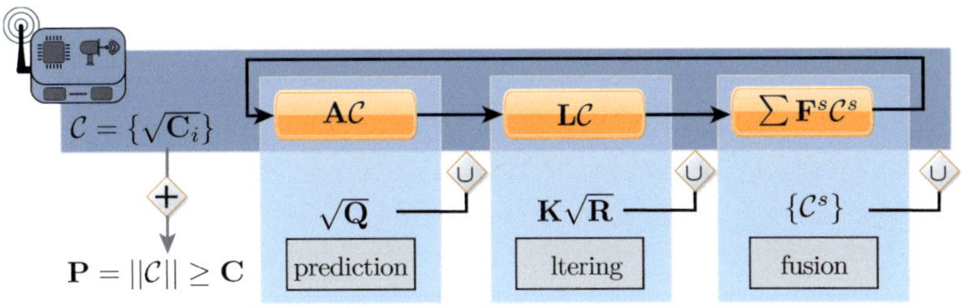

Figure 3.17: The processing of covariance square roots with the noise decomposition approach. All matrices are stored and processed separately in the set $\mathcal{C}^s$. In the fusion, the sets $\mathcal{C}^s$ are exchanged and the exact covariance is obtained by summing up the (squares of the) matrices.

small. However, the number of noise terms grows with every prediction and filter step as new process and measurement noise terms emerge. In particular, after the fusion, all noise terms comprised in at least one of the involved estimates must be processed. Therefore, even for short time horizons, computational and communication effort become considerable.

3.2.4 - b Bounded Reconstruction

In order to confine the effort at the sensors, an extension of the exact reconstruction technique for the aggregation of noise terms in residuals is proposed. To put it into another way, the idea is to process only "important" noise terms explicitly and to subsume the remaining ones in a covariance part that needs to be bounded in the fusion.

Let $\mathcal{I}_C^s \subseteq \mathcal{I}^s$ denote the indices of an arbitrary subset of noise terms that affect estimate $\underline{\hat{\mathbf{x}}}^s$. Then, the set

$$\mathcal{C}^s = \{(\sqrt{\mathbf{C}_i^s}, i) \mid i \in \mathcal{I}_C^s\} \tag{3.38}$$

contains all noise terms specified by $\mathcal{I}_C^s$ and defines the residual term

$$\mathbf{C}_b^s \geq \sum_{i \in \mathcal{I}^s \setminus \mathcal{I}_C^s} \sqrt{\mathbf{C}_i^s} \sqrt{\mathbf{C}_i^s}^\top . \tag{3.39}$$

The residual $\mathbf{C}_b^s$ is supposed to be a bound, which leads to the natural application of covariance bounds to the noise decomposition structure.

Definition 3.5 *The noise decomposition* $(\hat{\underline{\mathbf{x}}}, \mathcal{C}, \mathbf{C}_b)$ *is called* ***consistent*** *(for the state* $\underline{\mathbf{x}}$*) when*

$$\mathbf{C} = \mathrm{E}\{(\hat{\underline{\mathbf{x}}} - \underline{\mathbf{x}})^2\} \leq ||\mathcal{C}|| + \mathbf{C}_b .$$

As $\mathcal{C}$ gives precise information about the covariance and $\mathbf{C}_b$ bounds the remaining terms, a relation s established as follows.

Corollary 3.18 *Let* $(\hat{\underline{\mathbf{x}}}, \mathcal{C}, \mathbf{C}_b)$ *denote a consistent noise decomposition. Then*

$$\mathbf{P} = ||\mathcal{C}|| + \mathbf{C}_b \tag{3.40}$$

is a covariance bound.

In order to derive an algorithm that provides a consistent decomposition, the operations of linear estimators are examined. Prediction and filtering do not involve (partially unknown) cross-covariance matrices so that consistency follows directly with (3.18).

Lemma 3.19 *Let* $(\hat{\underline{\mathbf{x}}}, \mathcal{C}, \mathbf{C}_b)$ *be consistent for the state* $\underline{\mathbf{x}}$ *and let* i^v *and* i^w *denote unique identifiers for process and measurement noise terms of the linear system from Definition 2.1. Then,*

- $(\mathbf{A}\hat{\underline{\mathbf{x}}}, \mathbf{A}\mathcal{C} \cup \{(\sqrt{\mathbf{Q}}, i^w)\}, \mathbf{A}\mathbf{C}_b(\mathbf{A})^\top)$ *is consistent for the state* $\mathbf{A}\underline{\mathbf{x}} + \underline{\mathbf{w}}$ *and*

- $(\mathbf{L}\hat{\underline{\mathbf{x}}} + \mathbf{K}\underline{\mathbf{z}}, \mathbf{L}\mathcal{C} \cup \{(\mathbf{K}\sqrt{\mathbf{R}}, i^v)\}, \mathbf{L}\mathbf{C}_b(\mathbf{L})^\top)$ *with* $\mathbf{L} = (\mathbf{I} - \mathbf{K}\mathbf{H})$ *is consistent for the state* $\underline{\mathbf{x}}$.

According to Corollary 3.18, a noise decomposition is consistent, if the sum of two matrices bounds the true covariance. For this property, the allocation of noise terms into explicitly stored terms and residuals is irrelevant. In particular, it is feasible to reduce the number of stored terms without violating consistency.

Lemma 3.20 *Let* $(\hat{\underline{\mathbf{x}}}, \mathcal{C}, \mathbf{C}_b)$ *be consistent. Then* $(\hat{\underline{\mathbf{x}}}, \mathcal{C}\backslash\tilde{\mathcal{C}}, \mathbf{C}_b + ||\tilde{\mathcal{C}}||)$ *with* $\tilde{\mathcal{C}} \subseteq \mathcal{C}$ *is consistent.*

In the fusion operation, unknown correlations between the residuals must be bounded. To this end, the techniques from Section 3.2 are utilized. More precisely, cross-covariance matrices are reconstructed for those noise terms that are stored in all sets $\mathcal{C}^s$ while the remaining covariances are bounded.

Lemma 3.21 *Let* $(\hat{\underline{\mathbf{x}}}^s, \mathcal{C}^s, \mathbf{C}_b^s)$ *be consistent for* $s \in \mathcal{S}$ *and let* $\mathcal{I}_C^s$ *denote the indices of stored terms in* $\mathcal{C}^s$. *The sets*

$$\mathcal{C}_\cap^s := \{(\sqrt{\mathbf{C}_i^s}, i) \mid \sqrt{\mathbf{C}_i^s} \in \mathcal{C}^s \ , i \in \bigcap_{s\in\mathcal{S}} \mathcal{I}_C^s\} \subseteq \mathcal{C}^s$$

specify the intersection of noise terms. Then, $(\hat{\underline{\mathbf{x}}}, \mathcal{C}, \mathbf{C}_b)$ *with*

$$\hat{\underline{\mathbf{x}}} = \sum_{s\in\mathcal{S}} \mathbf{F}^s \hat{\underline{\mathbf{x}}}^s \tag{3.41}$$

$$\mathcal{C} = \Big\{\Big(\sum_{s\in\mathcal{S}} \mathbf{F}^s \sqrt{\mathbf{C}_i^s}, i\Big) \mid \sqrt{\mathbf{C}_i^s} \in \mathcal{C}_\cap^s\Big\} \tag{3.42}$$

$$\mathbf{C}_b = \sum_{s\in\mathcal{S}} \frac{1}{\omega^s} \mathbf{F}^s (\mathbf{C}_b^s + ||\mathcal{C}^s\backslash\mathcal{C}_\cap^s||)(\mathbf{F}^s)^\top \tag{3.43}$$

is consistent for $\omega^s > 0$, $\sum_{s\in\mathcal{S}} \omega^s = 1$ *and fusion gains* $\mathbf{F}^s$ *with* $\sum_{s\in\mathcal{S}} \mathbf{F}^s = \mathbf{I}$.

Proof. First, Lemma 3.20 is invoked to derive consistent noise decompositions $(\hat{\underline{\mathbf{x}}}^s, \mathcal{C}^s_\cap, \mathbf{C}^s_b + ||\mathcal{C}^s \backslash \mathcal{C}^s_\cap||)$. As discussed in Theorem 3.17, covariances are the sums of transformed noise matrices.

According to (3.3), error processes are transformed with $\mathbf{F}^s$ such that the overall noise transformations are given by $\sum_{s\in\mathcal{S}} \mathbf{F}^s \mathbf{T}^s_i$ with $\mathbf{T}^s_i = \mathbf{0}$, if $i \notin \mathcal{I}^s_C$. Hence, for the explicitly considered noise terms $\mathbf{C}^\nu_i$, $i \in \bigcap_{s\in\mathcal{S}} \mathcal{I}^s_C$, the square root representations $\sqrt{\mathbf{C}_i}$ must satisfy $\sum_{s\in\mathcal{S}} \mathbf{F}^s \mathbf{T}^s_i \sqrt{\mathbf{C}^\nu_i}$, which corresponds to the covariance transformation described in $\mathcal{C}$.

The (new) residuals $\mathbf{C}^s_b + ||\mathcal{C}^s \backslash \mathcal{C}^s_\cap||$ are inflated with the scalar factors from Lemma 3.12, yielding a joint covariance matrix bound, which, in turn, defines a bound on the true residual with Lemma 3.9. □

For an analysis of Lemma 3.21, consider the aggregated noise decompositions $(\hat{\underline{\mathbf{x}}}^s, \mathcal{C}^s_\cap, \mathbf{C}^s_b + ||\mathcal{C}^s \backslash \mathcal{C}^s_\cap||)$, where all noise terms that are not in the intersection $\bigcap_{s\in\mathcal{S}} \mathcal{I}^s_C$ are added to the residual terms. Then, analogies to other fusion techniques under unknown correlations can be drawn.

In contrast to split CI, the cross-covariance matrices defined by noise terms in $\mathcal{C}^s_\cap$ are in general different from zero. In particular, the set structure $\mathcal{C}$ is maintained through the fusion, which enables the explicit consideration of dependencies in subsequent operations. As already indicated in the notation, the residual term $\mathbf{C}^s_b + ||\mathcal{C}^s \backslash \mathcal{C}^s_\cap||$ corresponds to the covariance $\mathbf{C}^s_b$ from Section 3.2.3 and is inflated with the same technique to achieve a bound. In fact, in the proof of Lemma 3.21, the joint covariance matrices

$$\bar{\mathbf{C}}_a = \sum_{i\in\bigcap_{s\in\mathcal{S}} \mathcal{I}^s_C} \begin{pmatrix} \sqrt{\mathbf{C}^{s_1}_i}(\sqrt{\mathbf{C}^{s_1}_i})^\top & \dots & \sqrt{\mathbf{C}^{s_1}_i}(\sqrt{\mathbf{C}^{s_S}_i})^\top \\ \vdots & \ddots & \vdots \\ \sqrt{\mathbf{C}^{s_S}_i}(\sqrt{\mathbf{C}^{s_1}_i})^\top & \dots & \sqrt{\mathbf{C}^{s_S}_i}(\sqrt{\mathbf{C}^{s_S}_i})^\top \end{pmatrix} \tag{3.44}$$

and

$$\bar{\mathbf{P}}_b = \begin{pmatrix} \frac{1}{\omega^{s_1}}(\mathbf{C}_b^{s_1} + ||\mathcal{C}^{s_1} \backslash \mathcal{C}_\cap^{s_1}||) & \dots & \mathbf{0} \\ \vdots & \ddots & \vdots \\ \mathbf{0} & \dots & \frac{1}{\omega^{s_S}}(\mathbf{C}_b^{s_S} + ||\mathcal{C}^{s_S} \backslash \mathcal{C}_\cap^{s_S}||) \end{pmatrix} \quad (3.45)$$

are constructed implicitly such that $\bar{\mathbf{P}} = \bar{\mathbf{C}}_a + \bar{\mathbf{P}}_b$ constitutes a bound in the joint space. Indeed, the **bounding theory** from Section 3.2.2 is directly applicable to the joint covariance matrix bound. The fusion gains, for example, are obtained with Theorem 3.14.

Corollary 3.22 *Let* $\bar{\mathbf{P}} = \bar{\mathbf{C}}_a + \bar{\mathbf{P}}_b$ *be the joint covariance matrix bound obtained with* (3.44) *and* (3.45). *The fusion gains* $\mathbf{F} = (\mathbf{F}^{s_1} \dots \mathbf{F}^{s_S})$ *that minimize the bound of the fused estimate* $\mathbf{P} = ||\mathcal{C}|| + \mathbf{C}_b$ *from Lemma 3.21 are given by* $\mathbf{F} = \mathbf{P}\underline{\mathbf{1}}(\bar{\mathbf{P}})^+$.

Considerations concerning $\{\omega^s\}_{s \in \mathcal{S}}$ and convexity of the optimization from Theorem 3.15 also apply to the fusion formulas from Corollary 3.22. In terms of the α assessment from (3.32), the relation between $||\mathcal{C}_\cap^s||$ and $\mathbf{C}_b^s + ||\mathcal{C}^s \backslash \mathcal{C}_\cap^s||$ determines the tightness.

Now consider a linear estimator specified in Algorithm 2.1. The corresponding processing of $(\hat{\underline{\mathbf{x}}}, \mathcal{C}, \mathbf{C}_b)$ for unbiased linear estimators is summarized in Algorithm 3.3.

Algorithm 3.3 Sensor Processing of the Noise Decomposition

1: Initialization: $(\hat{\underline{\mathbf{x}}}, \mathcal{C}, \mathbf{C}_b) = (\hat{\underline{\mathbf{x}}}_0, \{(\sqrt{\mathbf{C}_0}, i_0^{x_s})\}, \mathbf{0})$
2: **for** $k = 1; k \in \mathcal{K}; k = k + 1$ **do**
3: Prediction: $(\mathbf{A}\hat{\underline{\mathbf{x}}}, \mathbf{A}\mathcal{C} \cup \{(\sqrt{\mathbf{Q}}, i_{k-1}^w)\}, \mathbf{A}\mathbf{C}_b(\mathbf{A})^\top)$
4: Filtering: $(\mathbf{L}\hat{\underline{\mathbf{x}}} + \mathbf{K}\underline{\mathbf{z}}, \mathbf{L}\mathcal{C} \cup \{(\mathbf{K}\sqrt{\mathbf{R}}, i_k^{v_s})\}, \mathbf{L}\mathbf{C}_b(\mathbf{L})^\top)$
5: Aggregation: $(\hat{\underline{\mathbf{x}}}, \mathcal{C} \backslash \tilde{\mathcal{C}}, \mathbf{C}_b + ||\tilde{\mathcal{C}}||)$ with $\tilde{\mathcal{C}} \subseteq \mathcal{C}$
6: Fusion: $(\sum_{s \in \mathcal{S}} \mathbf{F}^s \hat{\underline{\mathbf{x}}}^s$, (3.42), (3.43))
7: **end for**

The indices i are to be chosen such that they uniquely identify noise terms $\mathbf{C}_0^s$, $\mathbf{Q}_k$, and $\mathbf{R}_k^s$, $s \in \mathcal{S}$, $k \in \mathcal{K}$. Viable choices are for example $i_s^{x_0} = s$, $i_k^{v_s} = k \cdot (S+1) + s$, and $i_k^w = k \cdot (S+1) + S + 1$. Combining the consistency results for noise decompositions from Lemmata 3.19 to 3.21 with Algorithm 3.3 leads to the following statement about consistency.

Theorem 3.23 *Let the models be as in Definition 2.1 and let all nodes employ the linear estimator from Algorithm 3.3. For uncorrelated initial estimates* $\hat{\underline{\mathbf{x}}}_0^s$ *with* $\mathbf{C}_0^s \geq \mathrm{E}\{(\hat{\underline{\mathbf{x}}}_0^s - \underline{\mathbf{x}}_0)^2\}$, $s \in \mathcal{S}$,

$$\mathbf{P}^s = ||\mathcal{C}^s|| + \mathbf{C}_b^s \geq \mathbf{C}^s = \mathrm{E}\{(\hat{\underline{\mathbf{x}}}^s - \underline{\mathbf{x}})^2\} \tag{3.46}$$

is satisfied in all time steps.

PROOF. The claim is fulfilled at initialization according to the precondition. For the remaining operations, the claim has been proven in Lemmata 3.19 to 3.21. □

Hence, the noise decomposition approach is compatible with arbitrary linear estimators and provides covariance bounds that can be used for the optimization of filter and fusion gains. For an efficient application of the algorithm it is, however, necessary to aggregate noise terms. The associated trade-off between precision, computation, and communication becomes apparent in the next statement.

Theorem 3.24 *Let the assumptions be the same as in Theorem 3.23 and let* $\mathcal{A}_k^s$ *and* $\tilde{\mathcal{A}}_k^s$ *denote sets of noise terms that have been aggregated in time step* $k \in \mathcal{K}$ *at sensor* s *with two different noise selection policies. If* $\mathcal{A}_{\tilde{k}}^s \subseteq \tilde{\mathcal{A}}_{\tilde{k}}^s$, $\tilde{k} \leq k$, $s \in \mathcal{S}$, *it holds* $\mathbf{P}^s \leq \tilde{\mathbf{P}}^s$, *where* $\mathbf{P}^s$ *and* $\tilde{\mathbf{P}}^s$ *denote the bounds from Theorem 3.23 for the two policies.*

PROOF. For the sake of simplicity, all variables are defined only for the baseline policy with aggregation sets $\mathcal{A}_k^s$. Let analogous variables for the other aggregation policy be denoted by a tilde, e.g., $\mathbf{P}^s$ and $\tilde{\mathbf{P}}^s$. In a first step, the decomposition of the bound $\mathbf{P} = \mathbf{C}_a + \mathbf{C}_b$ is considered. It is proven for all operations that $\mathbf{C}_b^s \leq \tilde{\mathbf{C}}_b^s$ and $\mathbf{P}^s \leq \tilde{\mathbf{P}}^s$.

At initialization, the bounds are equal and the invariants are fulfilled. In prediction and filter steps, $\mathbf{C}_b$ and $\mathbf{P}$ are linearly transformed and noise terms $\mathbf{Q}$ and $\mathbf{KR}(\mathbf{K})^\top$ are added. As the operations are the same for both aggregation policies, the inequalities still hold according to Observation 3.8. In the aggregation step, the local bounds do not change. From $\mathcal{A}_k^s \subset \tilde{\mathcal{A}}_k^s$, it follows $||\mathcal{A}_k^s|| \leq ||\tilde{\mathcal{A}}_k^s||$, and, $\mathbf{C}_b \leq \tilde{\mathbf{C}}_b$. From $\mathbf{P} \leq \tilde{\mathbf{P}} \Leftrightarrow \mathbf{C}_a + \mathbf{C}_b \leq \tilde{\mathbf{C}}_a + \tilde{\mathbf{C}}_b$ it follows $\mathbf{C}_a - \tilde{\mathbf{C}}_a \leq \tilde{\mathbf{C}}_b - \mathbf{C}_b$. Now, consider the joint space covariance $\bar{\mathbf{C}}_a$ from (3.45) and the block diagonal matrix $\bar{\mathbf{C}}_b = \operatorname{blkdiag}(\mathbf{C}_b^{s_1}, \ldots, \mathbf{C}_b^{s_S})$. Let Δ denote the difference between the joint covariance matrices, i.e., $\Delta = \bar{\mathbf{C}}_a - \bar{\tilde{\mathbf{C}}}_a$. Then, the block diagonal entries of Δ are smaller than $\bar{\tilde{\mathbf{C}}}_b - \bar{\mathbf{C}}_b$. Therefore, $\Delta \leq \Delta(\omega)$, where $\Delta(\omega)$ denotes any bounding matrix obtained from $\bar{\tilde{\mathbf{C}}}_b - \bar{\mathbf{C}}_b$ with Lemma 3.12. However, that implies $\Delta \leq \bar{\tilde{\mathbf{P}}}_b - \bar{\mathbf{P}}_b$ with $\bar{\mathbf{P}}_b$ from (3.44). Therefore,

$$\bar{\mathbf{P}} = \bar{\mathbf{C}}_a + \bar{\mathbf{P}}_b = \bar{\tilde{\mathbf{C}}}_a + \Delta + \bar{\mathbf{P}}_b \leq \bar{\tilde{\mathbf{C}}}_a + \bar{\tilde{\mathbf{P}}}_b = \bar{\tilde{\mathbf{P}}} \, .$$

The invariants follow immediately with Lemma 3.9. □

According to Theorem 3.24, the precision of the noise decomposition approach **improves monotonically** with the consideration of additional noise terms. As already discussed above, CI is a special instance of the noise decomposition technique with $\mathcal{C} = \emptyset$, i.e., CI corresponds to Algorithm 3.3 when all noise terms are aggregated in each time step. Hence, the bounds provided by the noise decomposition approach are smaller than or equal to the ones obtained with CI, which means that CI defines the worst-case precision for the noise decomposition technique.

The other extreme is an exact reconstruction of the covariance that is achieved by storing all noise terms explicitly. The remaining cases necessitate noise selection policies.

3.2.4 - c Noise Selection Policies

The precision of the bound obtained by the noise decomposition approach is determined by the policy that is used in the aggregation step for the selection of noise terms. Consider for example a sensor network, where nodes exchange and combine estimates in a hop-to-hop communication. Then, the diffusion of measurement information through the network goes along with the distribution of noise terms. In fact, an estimate $\hat{\underline{\mathbf{x}}}_k^{s_1}$ is affected by measurement noise $\underline{\mathbf{v}}_{\tilde{k}}^{s_2}$ only when the estimate $\hat{\underline{\mathbf{x}}}_{\tilde{k}}^{s_2}$ (or a successor) has been communicated to sensor s_1 in the time period $\tilde{k}$ to k. Therefore, in order to optimize the bounds, the intersection $\bigcap_{s \in S} \mathcal{I}_C^s$ must be anticipated and maximized under consideration of the impact of individual noise terms on the estimation covariance.

Hence, the optimal noise selection technique involves the network structure and depends on the estimator processing. For the sake of applicability and simplicity, here, the focus is laid on general policies to manage the aggregation of noise terms locally. As a motivation, consider the drawbacks of a naïve selection policy that are illustrated in the next example.

Example 3.7: Selection Policy for the Noise Decomposition Approach

Example 3.3 is considered. All nodes are assumed to employ local KFs and the fusion technique from Theorem 3.14. For the reconstruction of cross-covariance matrices, noise terms in a varying number are stored explicitly. Let the noise terms be numbered consecutively such that measurement noise $\underline{\mathbf{v}}_k^s$ has the index $k \cdot 4 + s$, process noise $\underline{\mathbf{w}}_k$ is indexed with $k \cdot 4 + 4$, and the initial uncertainties are treated as $\underline{\mathbf{v}}_0^s$ with $s \in \{1, 2, 3\}$, $k \in \{0, \ldots, 9\}$.

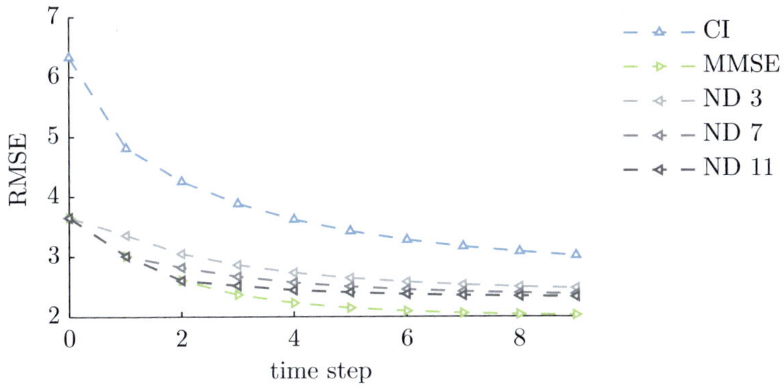

Figure 3.18: The RMSE of the bound of node s_3 for different noise selection policies. The bounds of the ND technique correspond to the MMSE result as long as the number of stored noise terms does not exceed the overall number of noise terms in the sensor network.

The utilized noise selection policy is to store only the latest noise terms ordered by their index. The RMSE of the obtained bounds for different numbers of noise terms is compared in Figure 3.18. Note that CI equals ND 0 and the MMSE approach that is implemented by means of an omniscient observer utilizes Theorem 2.5, which corresponds to ND ∞.

As initial covariances are bounded when CI is applied, the bound provided by CI is already worse than the ones obtained by the other approaches after the first fusion operation. The noise decomposition approach yields the MMSE bounds when the number of stored noise terms is larger than or equal to the number of noise terms that have emerged in the estimation process. Note that the number of noise terms has been chosen to cover all noise terms for specific time periods. When noise terms must be bounded, covariances are inflated and suboptimal bounds are achieved.

The explicit storage of noise terms leads to an improvement of the bounds compared to CI. However, this is only achieved by increasing the amount of processed and communicated data significantly.

The main reason is that many "empty" noise terms with $(\mathbf{0}, i)$ are processed. For example at initialization, it would be sufficient for the nodes to know that their own estimate is independent from the other ones. With the simple noise selection policy, three terms have to be considered at each node to ensure that the intersection $\bigcap_{s \in \mathcal{S}} \mathcal{I}_C^s$ from Lemma 3.21 entails all noise terms. ■

Although the noise decomposition clearly outperforms CI even for a few noise terms in Example 3.7, most of the stored noise terms are empty and do not improve the covariance bound. Indeed, as the sensor network consists of only three nodes, the effect is limited. However, in large-scale sensor networks, storing empty tuples wastes resources without improving the performance.

A remedy is to discard only the tuples of noise terms that are older than a time horizon $\bar{\tau}$. Then, all noise terms from time steps $k - \bar{\tau}$ to k are known in the fusion, as they are either stored in $\mathcal{C}^s$ or the **implicit knowledge** that $\sqrt{\mathbf{C}_i^s} = \mathbf{0}$ for $i \notin \mathcal{I}_C^s$ can be exploited. In terms of the fusion formula from Lemma 3.21, this policy is implemented by setting

$$\tilde{\mathcal{C}}^s = \mathcal{C}^s \cup \{(\mathbf{0}, i) \mid i \notin \mathcal{I}_C^s \,, k(i) \geq k - \bar{\tau}\} \,, \tag{3.47}$$

where $k(i)$ denotes the time step when noise term i has emerged. However, by using the time horizon as optimization parameter, the effort at the sensors is not precisely adjustable. As a matter of fact, the number of noise terms in $\mathcal{C}^s$ varies and is only loosely bounded by the number of sensors times the maximum time horizon $\bar{\tau}$.

Usually, the objective is to guarantee tight bounds while concurrently the communication effort in terms of the number of noise terms that need to be transmitted is restricted. In order to fulfill these conditions, a **prioritization algorithm** can be employed. Let $|\mathcal{C}|$ denote the number of noise terms in $\mathcal{C}$ and let $\bar{c}$ denote the maximum number of tuples to store in $\mathcal{C}$. With $\pi : \{1, \ldots, |\mathcal{C}|\} \to \{1, \ldots, |\mathcal{C}|\}$, a permutation is denoted that reflects an ordering of the entries in $\mathcal{C}$ according to

$$\operatorname{tr}\left\{\left(\sqrt{\mathbf{C}_{\pi(1)}}\right)^2\right\} \geq \cdots \geq \operatorname{tr}\left\{\left(\sqrt{\mathbf{C}_{\pi(|\mathcal{C}|)}}\right)^2\right\} \ .$$

Then, the noise aggregation with

$$\tilde{\mathcal{C}} = \{(\sqrt{\mathbf{C}_i}, i) \mid i \in \{\pi(\bar{c}+1), \ldots, \pi(|\mathcal{C}|)\}\} \tag{3.48}$$

dismisses the noise terms with the lowest impact on the estimation error. In order to exploit the partial independence of estimates in the fusion, noise terms are removed from $\mathcal{C}$ only when they are from time steps $\tilde{k} < k - \bar{\tau}$. Otherwise, the tuples $(\sqrt{\mathbf{C}_i^s}, i)$ are replaced by placeholders $(0, i)$ to indicate that the noise term has been removed. Then, implicit knowledge about the independence from (3.47) can be utilized again while additional storage and communication requirements caused by the placeholders are insignificant for high-dimensional systems.

Additionally, the split CI idea proposed in Section 3.2.3 can be used to combine multiple noise terms. For example, when node s does not exchange its estimate at time step k, the measurement noise terms of node s of time steps k and $k+1$ can be combined to form an **artificial noise term**. Another extension pertains to different requirements in computation and communication. In many scenarios, the costs for the latter are higher such that it can be appropriate to generate temporary noise decompositions $(\hat{\underline{\mathbf{x}}}, \mathcal{C}\backslash\tilde{\mathcal{C}}, \mathbf{C}_b + ||\tilde{\mathcal{C}}||)$ exclusively for the communication with neighboring nodes while concurrently the original noise decompositions are utilized in subsequent local operations. A simple implementation of a more elaborate noise selection policy is examined in the next example.

Example 3.8: Optimized Noise Decomposition

Consider Example 3.7 again. The idea is to reduce computational and communication effort by utilizing the concept of implicit knowledge. To this end, let nodes store all noise terms that constitute the local estimation error and that do not reach back more than a particular number of time steps into the past. Then, the square roots of covariances larger than zero are stored explicitly and the remaining terms of the considered time steps are reconstructed by means of (3.47).

In order to achieve the MMSE results for the first 1, 2, or 3 time steps, it is necessary to store 3, 7, or respectively 11 noise terms with the naïve selection policy. Now consider the implementation of the implicit knowledge policy with $\bar{\tau} = 1, 2, 3$. Only those noise terms that are depicted in Table 3.1 must be stored and communicated. The remaining terms have no impact on the estimation error, i.e., they are implicitly known to be zero. In order to quantify communication requirements, the number of noise terms before the fusion must be considered. From Table 3.1 these numbers follow as 1, 4(5), or 8(9) for the different time horizons when implicit knowledge is exploited.

	Node s_1	Node s_2	Node s_3
$k=0$ (init)	$\underline{\mathbf{v}}_0^{s_1}$	$\underline{\mathbf{v}}_0^{s_2}$	$\underline{\mathbf{v}}_0^{s_3}$
$k=0$ (fus)	$\cdots, \underline{\mathbf{v}}_0^{s_3}$	$\cdots, \underline{\mathbf{v}}_0^{s_3}$	$\cdots, \underline{\mathbf{v}}_0^{s_1}, \underline{\mathbf{v}}_0^{s_2}$
$k=1$ (pred)	$\cdots, \underline{\mathbf{w}}_0$	$\cdots, \underline{\mathbf{w}}_0$	$\cdots, \underline{\mathbf{w}}_0$
$k=1$ (filt)	$\cdots, \underline{\mathbf{v}}_1^{s_1}$	$\cdots, \underline{\mathbf{v}}_1^{s_2}$	$\cdots, \underline{\mathbf{v}}_1^{s_3}$
$k=1$ (fus)	$\cdots, \underline{\mathbf{v}}_1^{s_3}$	$\cdots, \underline{\mathbf{v}}_1^{s_3}$	$\cdots, \underline{\mathbf{v}}_1^{s_1}, \underline{\mathbf{v}}_1^{s_2}$
$k=2$ (pred)	$\cdots, \underline{\mathbf{w}}_1$	$\cdots, \underline{\mathbf{w}}_1$	$\cdots, \underline{\mathbf{w}}_1$
$k=2$ (filt)	$\cdots, \underline{\mathbf{v}}_2^{s_1}$	$\cdots, \underline{\mathbf{v}}_2^{s_2}$	$\cdots, \underline{\mathbf{v}}_2^{s_3}$

Table 3.1: Noise terms that determine the estimation error in Example 3.7

Although the number of processed and transmitted noise terms is already smaller than for the naïve technique, the full potential of implicit knowledge is developed in large sensor networks with a dense communication. In those sensor networks, the communication of estimates takes several time steps, which leads to many zero entries in the noise decomposition of local estimation errors that can be neglected by means of implicit knowledge. ■

The proposed policies are supposed to provide general ideas that reduce computational and communication effort compared to the naïve solution. Indeed, other approaches are conceivable and techniques that are more efficient can be derived for specific network topologies. However, even the simple policies permit the calculations of tight bounds with limited capacities.

3.3 Conclusion

Processing covariances in sensor networks proves challenging as the precise quantification of dependencies is required to determine uncertainties of fused estimates. Dependencies, however, are determined by filter and fusion gains of all sensors so that they can be calculated only when transformations of remote sensors are known. A minimum requirement for the availability of such information is global model knowledge. Indeed, reasons to maintain covariances are compelling; meaningful optimizations of filter and fusion gains are based on them and for the possible subsequent utilization of estimates in control or decision problems, covariances provide indispensable quality information. The objective of this chapter was to provide covariances of estimates in sensor networks and to derive fusion methods premised on the assumption that no global model knowledge is available.

3.3.1 Sample-based Covariance Estimation

One approach to obtain covariances in distributed and decentralized estimation is to recursively process variables that allow the approximate description of dependencies between estimators. The idea pursued in Section 3.1 was to estimate (cross-)covariance matrices based on **sample representations** of error distributions. For this purpose, a sample set of the joint space error that depicts the dependencies between all estimators was considered. It turned out that the processing of these joint space samples can be decomposed into local operations.

The associated research questions narrowed down to deriving an efficient covariance estimator and to determining the distribution for the generation of samples that minimizes the error of covariance estimation. It was demonstrated that utilizing known means to calculate covariance estimates improves the precision. The **optimal sampling distribution** for scalar systems was identified to be a modified Bernoulli distribution, which even enables the determination of individual noise variances from samples without stochastic errors. The application of the results to general systems was discussed and the corresponding multivariate Bernoulli distribution was shown to optimize estimates of noise covariances.

The proposed sample-based covariance estimation with adjustable noise generation bears the potential to be applied in various scenarios. In particular, the technique is not confined to linear models and can easily be generalized to higher order moments. Thus, future research should continue towards methods for efficient sampling in multivariate systems subject to finite moments. Either, this can encompass derivations of alternative sampling distributions or further analysis of covariance estimators in combination with Bernoulli-distributed samples.

It could be possible to achieve considerable improvements in terms of the precision of covariance estimation by deterministic sampling methods. However, in the current setup, it is necessary to generate sample lists that are independent from an arbitrary number of other noise terms. This in turn, necessitates deterministic sampling schemes that provide infinitely many orthogonal sets. As these sets determine a minimum dimension of the sample space, no finite-dimensional deterministic scheme that provides error-free covariance estimates can be found. Still, it is conceivable that deterministic elements in the generation of samples yield approximately orthogonal sample sets with a higher precision than purely stochastic covariance estimates.

3.3.2 Bounding Techniques

The strategy pursued in the second part of this chapter was to process **bounds** as substitutes for covariances, which allows bypassing the explicit reconstruction of cross-covariance matrices. In Section 3.2.1, it was proven that the bound obtained in the fusion of two estimates with covariance intersection (CI) exhibits optimality subject to various criteria. Therefore, CI constitutes the method of choice for the fusion under unknown correlations. Thus, batch CI for the fusion of more than two estimates was examined and **generalized**. In this context, fundamental properties such as convexity of a necessary optimization were proven and the systematic inclusion of knowledge about cross-covariance matrices in the fusion was facilitated.

The derived techniques found application to linear systems in the ensuing section. A **decomposition of local covariances** into independent terms was proposed, and generalized CI was applied to the decomposition. It turned out that exploiting partially known cross-covariance matrices strictly improves the precision of bounds compared to (basic) CI. In fact, generalized CI combined with the noise decomposition of covariances constitutes an efficient algorithm for calculating bounds in sensor networks. Tightness of the obtained bound is adjustable between exact (cross-)covariance matrices and CI bounds subject to computational and communication effort.

Indeed, bounding theory is not exhausted with the results presented in this thesis. The rigorous examination of the fusion of more than two estimates under unknown correlations is still missing. In particular, the optimality statement for two estimates derived in this thesis, remains to be generalized. Another interesting research direction constitutes the application of the bounding theory to consensus techniques [126]. Spadework has been carried out in [12] by aiming to reach a consensus in the information form, which corresponds to a repeated application of covariance intersection without a weight optimization. Extending these results to varying weights certainly yields an improvement in precision even though the analysis is aggravated.

Further research is also needed to apply the noise decomposition technique efficiently. So far, only general policies have been proposed to **select noise covariances**. Selection policies that are adapted to typical scenarios promise a better exploitation of noise terms, which translates into tighter bounds. In an attempt to lower computational and communication effort of bounding techniques and still exploit knowledge about cross-covariance matrices, it is also advisable to apply the scalar correlation constraints from [67, 139]. If restricting cross-covariance matrices by means of scalar or low-dimensional vectors is possible, the need to store noise terms explicitly could be omitted.

CHAPTER 4

Hypothesizing Distributed Kalman Filtering

Sensor network theory is applied to various research areas that are concerned with properties of and interconnections between logically separated units. A particularly interesting field of application constitutes the **surveillance of large areas**. In the hope of achieving higher resolution and better coverage than it is realizable with a few (top-notch) devices, sensors are spatially distributed over an area to observe the phenomenon or target of interest at close range.

Practical examples for such systems are manifold. For surveillance of air space, several radar stations observe overlapping sectors, preprocess huge amounts of raw sensor data locally, and communicate tracking information to aerial surveillance control centers for further evaluation. A very different type of sensor networks are interconnected minicomputers equipped with simple sensing devices that perform basic tasks subject to limited computational load and energy resources. Such "smart dusks" [92] can for example be used to detect wildfires or monitor air pollution in cities.

Both examples feature nodes with computational capabilities to preprocess measurements. As a matter of fact, the availability of processing logic is inherently linked to applying nontrivial communication protocols, which require some – at least hard-coded – logic themselves. Indeed, computational tasks and capabilities of sensors vary considerably between simple threshold functions in smart dusks and elaborate tracking algorithms in radars. Still, in both scenarios, information is preprocessed at the sensors and the final evaluation is carried out in a control center.

In order to efficiently process information in such systems, communication between nodes is to be minimized and the estimate at the fusion center is to be optimized. In this chapter, linear **distributed estimation** with smart sensors is examined and proper estimators are developed. For this purpose, initially, a controlled environment is assumed where (stochastic) properties of the communication network are known, data exchanges are guided by the estimator, and sensor as well as fusion center processing can be optimized. Then, knowledge requirements about the parameters are successively reduced until only individually operating sensors aim to provide the fusion center as much information as possible.

For the illustration of key problems associated with distributed estimation in sensor networks, a centralized KF is compared to approaches that are based on local KFs.

Example 4.1: Local versus Global Filter Optimization

Consider an artificial scalar system defined by $\mathbf{A} = 1$, $\mathbf{Q} = 1$, and two sensors $s \in \{s_1, s_2\}$ with measurement models $\mathbf{H}_1^s = 1$ and $\mathbf{R}_1^s = 1$. The initial estimates at the sensors are denoted as $\hat{\underline{\mathbf{x}}}_0^s$ with variances $\mathbf{C}_0^s = 1$. When local KFs are applied at the sensors, the estimates after one prediction and one filter step are given (after some algebraic simplifications) by $\hat{\underline{\mathbf{x}}}_1^s = \frac{1}{3}\hat{\underline{\mathbf{x}}}_0^s + \frac{2}{3}\underline{\mathbf{z}}_1^s$. When initial estimates and measurements are processed with a centralized KF, the LMMSE estimate is obtained as

$$\hat{\underline{\mathbf{x}}}_1 = \frac{1}{8}\hat{\underline{\mathbf{x}}}_0^{s_1} + \frac{1}{8}\hat{\underline{\mathbf{x}}}_0^{s_2} + \frac{3}{8}\underline{\mathbf{z}}_1^{s_1} + \frac{3}{8}\underline{\mathbf{z}}_1^{s_2} \ .$$

Now, consider the fusion of local tracks $\hat{\underline{\mathbf{x}}}_1^s$. As the fused estimate is a linear combination of $\hat{\underline{\mathbf{x}}}_1^{s_1}$ and $\hat{\underline{\mathbf{x}}}_1^{s_2}$, the **inner ratios** between initial estimates and measurements cannot be modified. However, as they are $\frac{1}{3}$ to $\frac{2}{3}$ for local estimators and $\frac{1}{8}$ to $\frac{3}{8}$ for the centralized KF, the LMMSE estimate cannot be obtained from the locally calculated estimates anymore. In fact, the LMMSE combination from Theorem 2.5 yields

$$\underline{\tilde{\mathbf{x}}}_1 = \frac{1}{6}\underline{\hat{\mathbf{x}}}_0^{s_1} + \frac{1}{6}\underline{\hat{\mathbf{x}}}_0^{s_2} + \frac{2}{6}\underline{\mathbf{z}}_1^{s_1} + \frac{2}{6}\underline{\mathbf{z}}_1^{s_2} \ .$$

Simple calculations show that the MSE of $\underline{\tilde{\mathbf{x}}}_1$ is 3.4% higher than that of $\underline{\hat{\mathbf{x}}}_1$. Note that for more than two sensors or for a processing over several time steps, the MSE difference can be considerable as illustrated in [178]. ■

Example 4.1 demonstrates that the precision of the estimate at the fusion center depends on the filter processing at the sensors. In fact, it is a consequence of the application of **locally** optimal KFs that the **global** estimate is suboptimal. In an attempt to achieve the LMMSE estimate at the fusion center, the filter processing at the sensors is to be adapted to the LMMSE filter gains.

To this end, filter transformations of the centralized KF can be decomposed and carried out at the sensors as it has been proposed with the distributed Kalman filter (DKF) [59]. Then, only auxiliary (suboptimal) estimates are maintained at the sensors but the fusion result exhibits optimality. Now, the aim of this chapter is to illuminate **distributed estimation subject to limited knowledge**. First steps are made in Section 4.1 by deriving techniques to detect and cope with a bias that is induced in the decomposition scheme due to sensor failures. Indeed, a reliable deterministic communication and global measurement model knowledge are still required to apply the technique.

The generalization of the DKF to sensor networks with packet delays and losses is proposed in Section 4.2. Motivated by surveillance systems that aim to achieve an approximately constant coverage of a large area, the idea is to identify the measurement capacity of the sensor network, to estimate this variable, and to use a hypothesis for the optimization of filter gains at the sensors. In this way, the filter processing becomes independent from global model knowledge.

Still, a common hypothesis is necessary for the application of the so-called hypothesizing distributed Kalman filter. This condition is relaxed in Section 4.3 by considering individual hypotheses at the sensors. The concept is to optimize filter gains according to locally available best guesses about the measurement capacity of the sensor network. In this context, covariance bounds and an optimized fusion method are derived based on the results of Chapter 3, which are also applicable to basic hypothesizing filtering.

4.1 Decomposition of the Central Kalman Filter

The LMMSE estimator for linear systems is given by a central KF. In the context of sensor networks, this insight has already been exploited in centralized estimation, where sensors communicate measurements to the fusion center and a central KF is employed for the combination. In this section, the concept is applied to distributed estimation.

For this purpose, consider the operations of a centralized KF that processes several measurements at a particular time step, predicts the estimate, and continues with measurements of the next time step. As all models and transformations are linear, filtering an individual measurement and predicting it to the current time step can be construed as an isolated operation. Indeed, the estimate of a centralized KF is the sum of all these "measurement packages". Now, decomposition approaches allot the calculation and combination of measurement packages to the sensor sources, as it is illustrated in Figure 4.1.

While the processing in prediction and filtering is defined by the centralized KF equations, degrees of freedom exist for the realization

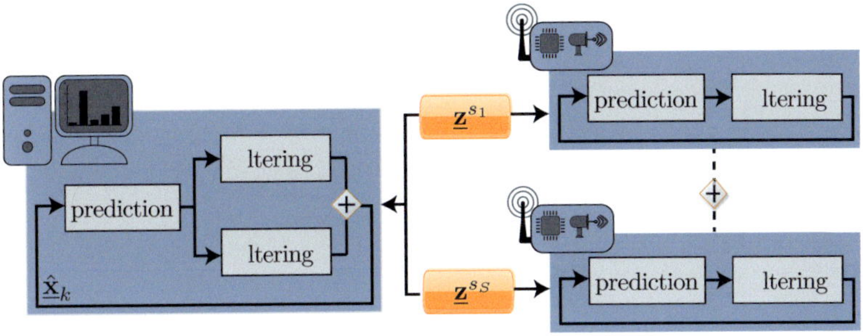

Figure 4.1: The calculations of a centralized KF boil down to the summation of linearly transformed measurements. The aggregation over several time steps is carried out at the sensors.

in distributed estimation. In particular, it is viable to implement different strategies to fuse data. For example, either data packages consisting of measurements since the last (successful) transmission or packets consisting of measurements from all time steps can be transmitted to the fusion center. Moreover, the fusion center can either fuse received data packages into a local estimate or replace previously received information.

In the following, measurements are recursively processed at the sensors and data packages are replaced by the latest information from sensors. More precisely, a decomposition of the centralized KF is derived and the following techniques are presented:

- Sensor and fusion center processing of the **distributed Kalman filter** are examined.
- A technique to detect sensor and communication failures based on an **additive bias** term is derived.
- A technique to correct biased estimators with locally calculated **multiplicative debiasing matrices** is proposed.

The results of this section serve as the mathematical framework for the (advanced) estimators presented in subsequent sections. In particular, the decomposition of operations and the optimization of filter operations according to a centralized estimator are key components of efficient estimators for distributed systems.

4.1.1 Distributed Kalman Filter

The DKF is the natural **decomposition of a centralized KF** that has been originally proposed by exploiting the conditional independence of measurements in the context of Gaussian densities [56,59,60, 100]. The concept is to combine transformed measurements from the same source in one vector so that only few data must be transmitted to the fusion center to generate the central KF estimate.

The following derivation uses the LMMSE criterion to obtain a linear estimator that fits to the structure of Algorithm 2.1. It simplifies the original formulas [59] by freeing the equations from a scaling factor. As already mentioned, the idea is to **produce tracks from measurements** at the sensors, transmit the tracks to the fusion center, and to generate an estimate from available data when needed.

Such a "replacement strategy" may appear counterintuitive since a simpler approach would be to fuse all received data into a recursively held estimate. However, assume that sensors regularly transmit the latest estimate and the fusion center combines them with its predicted estimate. As the estimates from the sensors contain all (local) measurements from previous time steps, measurements are included into the estimate at the fusion center multiple times. More precisely, when a sensor transmits its estimate in every time step, the measurement from the last time step is included twice, the one from two time steps ago thrice, and so on. In this context, it proves difficult to achieve filter gains that are equivalent to those of the central KF.

The central KF processing is specified in Section 2.2.2. An algebraically equivalent form for the calculation of filter gains is obtained by applying the Woodbury matrix inversion lemma [171]. Let $\mathcal{M}_k$ denote indices of measurements observed at time step k. The **information form** of the central KF [116] for the processing of several measurements is determined by gains

$$\mathbf{L}_k = \mathbf{C}_{k|k}(\mathbf{C}_k)^{-1} \text{ and } \mathbf{K}_k^m = \mathbf{C}_{k|k}(\mathbf{H}_k^m)^\top(\mathbf{R}_k^m)^{-1} , \tag{4.1}$$

where the posterior covariance is obtained as

$$\mathbf{C}_{k|k} = \Big((\mathbf{C}_k)^{-1} + \sum_{m\in\mathcal{M}_k} (\mathbf{H}_k^m)^\top(\mathbf{R}_k^m)^{-1}\mathbf{H}_k^m\Big)^{-1} . \tag{4.2}$$

Now, consider a network where all sensors are equipped with computing power and assume that global model knowledge is available. The sensor estimates are denoted by $\hat{\underline{\mathbf{x}}}_k^s$ and the estimate at the fusion center by $\hat{\underline{\mathbf{x}}}_k$. The processing of the DKF is derived based on the premise that $\sum_{s\in\mathcal{S}} \hat{\underline{\mathbf{x}}}_k^s$ provides the central KF result at every processing step of Algorithm 2.1.

Let the same initial estimate $(\hat{\underline{\mathbf{x}}}_0, \mathbf{C}_0)$ be available to all sensors at **initialization**. Then, local estimators are initialized with $\mathbf{T} = \frac{1}{|\mathcal{S}|}\mathbf{I}$ and $\hat{\underline{\mathbf{x}}}_0^s = \hat{\underline{\mathbf{x}}}_0$. If independent estimates $(\tilde{\hat{\underline{\mathbf{x}}}}_0^s, \tilde{\mathbf{C}}_0^s)$ are provided, the LMMSE fusion is the convex combination of the estimates and the sum structure is established with $\mathbf{T}^s = \mathbf{C}_0(\mathbf{C}_0^s)^{-1}$, where $\mathbf{C}_0 = (\sum_{s\in\mathcal{S}}(\mathbf{C}_0^s)^{-1})^{-1}$.

In the **prediction** step, the standard formulas for linear estimators (2.21) and (2.22) are applied, leading to

$$\hat{\underline{\mathbf{x}}}_{k+1}^s = \mathbf{A}_k\hat{\underline{\mathbf{x}}}_k^s , \tag{4.3}$$

$$\mathbf{C}_{k+1} = \mathbf{A}_k\mathbf{C}_k(\mathbf{A}_k)^\top + \mathbf{Q}_k . \tag{4.4}$$

For the **filtering** of measurements, the filter gains of a central KF in the information form (4.1) are used. When one measurement is observed per sensor and time step, the processing at an arbitrary sensor s amounts to

$$\hat{\underline{\mathbf{x}}}^s_{k|k} = \mathbf{L}_k \hat{\underline{\mathbf{x}}}^s_k + \mathbf{K}^s_k \underline{\mathbf{z}}^s_k \,, \tag{4.5}$$

$$\mathbf{C}_{k|k} = \Big((\mathbf{C}_k)^{-1} + \sum_{s\in\mathcal{S}} (\mathbf{H}^s_k)^\top (\mathbf{R}^s_k)^{-1} \mathbf{H}^s_k\Big)^{-1} \,, \tag{4.6}$$

where filter gains are obtained as

$$\mathbf{L}_k = \mathbf{C}_{k|k}(\mathbf{C}_k)^{-1} \text{ and } \mathbf{K}^s_k = \mathbf{C}_{k|k}(\mathbf{H}^s_k)^\top (\mathbf{R}^s_k)^{-1} \,. \tag{4.7}$$

Note that the **global covariance** $\mathbf{C}_k$ is processed at the sensors instead of the true covariances of the estimates, which are specified by $\mathbf{C}^s_k = \mathrm{E}\{(\hat{\underline{\mathbf{x}}}^s_k - \underline{\mathbf{x}})^2\}$. This is necessary as the filter processing and, in particular, the gains in (4.7) are optimized with the global covariance instead of local covariances. As the formulas were derived on the premise that the sum $\sum_{s\in\mathcal{S}} \hat{\underline{\mathbf{x}}}^s_k$ equals the central KF estimate, the following statement is obtained.

Theorem 4.1 *Let* $\hat{\underline{\mathbf{x}}}_0 = \sum_{s\in\mathcal{S}} \hat{\underline{\mathbf{x}}}^s_0$ *correspond to the central KF estimate with covariance* $\mathbf{C}_0$ *and let* $\hat{\underline{\mathbf{x}}}^s_k$ *denote estimates obtained with the DKF processing from* (4.3) *and* (4.5)*, i.e., by a linear estimator from Algorithm 2.1 with filter gains* (4.7)*. Then,*

$$\hat{\underline{\mathbf{x}}}_k = \sum_{s\in\mathcal{S}} \hat{\underline{\mathbf{x}}}^s_k \tag{4.8}$$

is the LMMSE estimator and $\mathbf{C}_k = \mathrm{E}\{(\hat{\underline{\mathbf{x}}}_k - \underline{\mathbf{x}})^2\}$.

Note that the DKF is a strict implementation of the decomposition concept illustrated in Figure 4.1. In fact, sensors apply the central KF transformations to locally observed measurements. The aggregation of measurements leads to estimates $\hat{\underline{\mathbf{x}}}^s_k$ that correspond to the sum of preprocessed measurement packages of one source and entail all information collected at sensor s up to time step k.

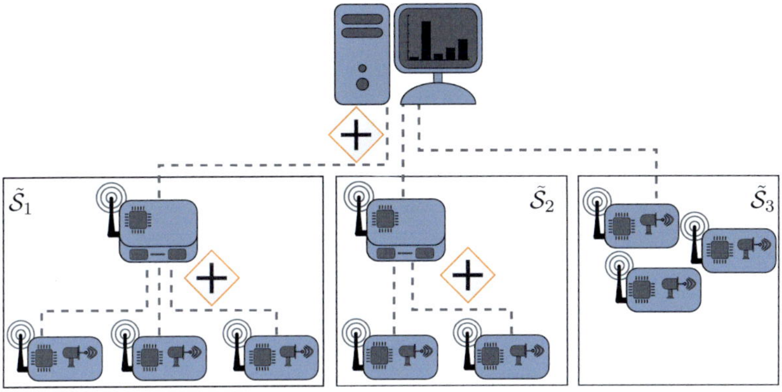

Figure 4.2: A hierarchical sensor network with in-network processing. Due to the additive structure of the centralized KF, estimates from children can be aggregated during communication.

It is also worth pointing out the **additive structure** of the DKF. As the same transformation matrices $\mathbf{L}_k$ and $\mathbf{A}_k$ are applied at all sensors, measurement terms $\mathbf{K}_k^s \underline{\mathbf{z}}_k^s$ can be added to arbitrary estimates $\hat{\underline{\mathbf{x}}}_k^{\tilde{s}}$ without affecting the result from Theorem 4.1. In particular, it is possible to process several measurements at one sensor per time step and to aggregate estimates in a hop-to-hop communication to the fusion center. Consider for example hierarchical sensor networks where at each sensor estimates $\hat{\underline{\mathbf{x}}}_k^{\tilde{s}}$ from children $\tilde{s} \in \tilde{\mathcal{S}} \subset \mathcal{S}$ are summed up and only the aggregated vector $\hat{\underline{\mathbf{x}}}_k^{\tilde{\mathcal{S}}} = \sum_{\tilde{s} \in \tilde{\mathcal{S}}} \hat{\underline{\mathbf{x}}}_k^{\tilde{s}}$ is transmitted to the fusion node, as it is illustrated in Figure 4.2. Such an **in-network-processing** can significantly reduce communication costs without affecting the precision of the fused estimate.

Indeed, for the calculation of KF gains, the global covariance $\mathbf{C}_k$, which depends on the **measurement capacity**

$$\mathbf{C}_k^z := \sum_{s \in \mathcal{S}} (\mathbf{H}_k^s)^\top (\mathbf{R}_k^s)^{-1} \mathbf{H}_k^s \ , \tag{4.9}$$

is required. Hence, sensors need access to all measurement models, i.e., global measurement model knowledge according to Definition 2.5 must be available. However, unexpected sensor failures alter the (underlying) measurement capacity of the sensor network while functioning sensors still operate on the original value of $\mathbf{C}_k^z$ as long as the failure is not communicated to all sensors. This degrades the performance of the DKF as illustrated in the next example.

Example 4.2: Effect of Sensor Failures on Decomposition Scheme

A target that moves with constant velocity on a one-dimensional manifold, e.g., on a street, is considered. Let T denote a time interval and ps the power spectral density of the (white) noise in continuous time. Then, a discrete-time constant velocity model is derived from the continuous time counterpart according to [106] as

$$\underline{\mathbf{x}}_{k+1} = \begin{pmatrix} 1 & T \\ 0 & 1 \end{pmatrix} \underline{\mathbf{x}}_k + \underline{\mathbf{w}}_k \text{ with } \underline{\mathbf{w}} \sim \mathcal{N}(\underline{0}, \mathbf{Q}) \ , \tag{4.10}$$

where

$$\mathbf{Q} = ps \cdot \begin{pmatrix} T^3/3 & T^2/2 \\ T^2/2 & T \end{pmatrix} \ .$$

The target is tracked at a fusion center that receives data from two sensors that measure the position of the state according to

$$\underline{\mathbf{z}}_k^s = (1 \quad 0)\, \underline{\mathbf{x}}_k + \underline{\mathbf{v}}_k \text{ with } \underline{\mathbf{v}}_k \sim \mathcal{N}(0, 10),\ s {\in} \{s_1, s_2\} \ .$$

Both sensors are initialized with independent estimates subject to an uncertainty of $\mathbf{C}_0^s = 100 \cdot \mathbf{I}$ and the tracking process lasts 10 time steps with $T = 0.05$ and $ps = 10$. The target starts at position 10 with initial velocity 5. The sensors communicate with the fusion center at every time step but due to an erroneous communication device, only zero vectors are received from sensor s_2 after time step 5.

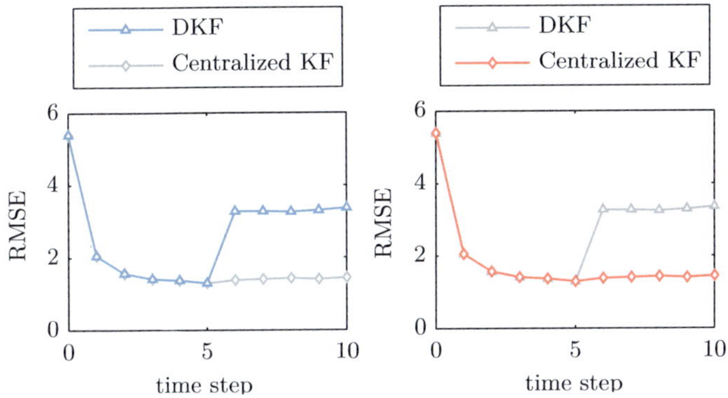

Figure 4.3: RMSE of the fusion center estimate in 100 Monte Carlo runs for DKF and centralized KF. After 5 time steps sensor s_2 fails and causes the estimate of the DKF to be biased.

For an evaluation, centralized KF and DKF are compared. In contrast to the presented version of the DKF, the implementation of [60] is utilized that aims to mitigate the bias of local estimates by scaling local estimates with the number of sensors. The centralized KF is assumed to use a gating procedure that detects and ignores zero vectors.

As it is demonstrated in Figure 4.3, the effect of sensor failures is striking. In the time period $k = [0, \ldots, 5]$, both estimators yield the same (LMMSE) results. This is a direct consequence of Theorem 4.1. However, while the estimation error of the centralized KF increases slightly due to missing measurements from sensor s_2, the RMSE of the DKF jumps within one time step by more than 100% as soon as communication failures produce an effect. ■

In Example 4.2, the main drawback of the DKF has been demonstrated. As the sensor processing of the DKF is exclusively optimized according to a central KF scheme, it holds $\mathbf{L}_k \neq \mathbf{I} - \mathbf{K}_k^s \mathbf{H}_k^s$. Thus, biased estimates with $\mathrm{E}\{\hat{\underline{\mathbf{x}}}_k^s\} \neq \mathrm{E}\{\underline{\mathbf{x}}\}$ are calculated at the sensors.

If all data is received at the fusion center, the summation in (4.8) eliminates this bias. However, if communication or sensor failures occur, the estimator at the fusion center is biased, which amounts to an unnecessarily high estimation error according to Lemma 2.2.

4.1.2 Error Reduction with Debiasing Approaches

As it has been discussed in Section 2.3, sensor networks are typically subject to imperfect communication and error-prone sensors. Hence, techniques are necessary to provide efficient and reliable estimates at the fusion center even if assumptions about the processing are not perfectly matched. According to Lemma 2.2, unbiased counterparts can be derived for biased estimators that yield a lower MSE. Thus, the remainder of this section is devoted to the derivation of techniques to either detect or correct the bias of sensor network estimators.

In linear estimation, the **(random) deviation** from the estimate to the true state can be traced. For this purpose, consider a linear estimator according to Algorithm 2.1. Then, the definition $\underline{\mathbf{b}} = \hat{\underline{\mathbf{x}}} - \underline{\mathbf{x}}$ leads to the following recursive formulas for biased estimators[1]

$$\text{Initialization: } \underline{\mathbf{b}}_0 = \hat{\underline{\mathbf{x}}}_0 - \underline{\mathbf{x}}_0 \ , \tag{4.11}$$

$$\text{Prediction: } \underline{\mathbf{b}}_{k+1} = \mathbf{A}_k \underline{\mathbf{b}}_k + \underline{\mathbf{w}}_k \ , \tag{4.12}$$

$$\text{Filtering: } \underline{\mathbf{b}}_{k|k} = \mathbf{L}_k \underline{\mathbf{b}}_k + (\mathbf{L}_k + \mathbf{K}_k \mathbf{H}_k - \mathbf{I}) \underline{\mathbf{x}}_k + \mathbf{K}_k \underline{\mathbf{v}}_k \ , \tag{4.13}$$

$$\text{Fusion: } \underline{\mathbf{b}}_k = \sum_{s \in \mathcal{S}} \mathbf{F}_k^s \underline{\mathbf{b}}_k^s + \Big(\sum_{s \in \mathcal{S}} \mathbf{F}_k^s - \mathbf{I} \Big) \underline{\mathbf{x}}_k \ . \tag{4.14}$$

The bias of an estimator is defined as the expected deviation to the true state.

[1] As a matter of fact, the definition of $\underline{\mathbf{b}}$ corresponds to the one of the error $\underline{\mathbf{e}}$ from Section 3.1.1. However, $\underline{\mathbf{e}}$ was defined in the context of unbiased estimators so that the (more general) formulas for $\underline{\mathbf{b}}$ are different in fusion and filtering steps.

Definition 4.1 *Let $\underline{\hat{\mathbf{x}}}$ denote a (potentially) biased estimator. The expectation* $\mathrm{E}\{\underline{\mathbf{b}}\}$ *with* $\underline{\mathbf{b}} = \underline{\hat{\mathbf{x}}} - \underline{\mathbf{x}}$ *is referred to as the* ***bias*** *of* $\underline{\hat{\mathbf{x}}}$.

The aim of **debiasing techniques** is to provide quantities that permit the elimination of the bias of an estimator, i.e., to derive the unbiased counterpart to a biased estimator. This is, in particular, relevant for the DKF from the last section but is also applicable to arbitrary linear estimators in distributed and decentralized estimation.

4.1.2 - a Additive Bias

In the first instance, the bias is quantified by means of additive vectors. A stochastically motivated idea is to estimate the bias $\underline{\mathbf{b}}$ and to apply a correction by means of $\underline{\hat{\mathbf{x}}} - \underline{\hat{\mathbf{b}}}$. To this end, it is sufficient to replace the unknown state in (4.11) to (4.14) by $\underline{\tilde{\mathbf{x}}} = \underline{\hat{\mathbf{x}}} - \underline{\hat{\mathbf{b}}}$. Then, it is easy to verify that the estimator $\underline{\tilde{\mathbf{x}}}$ is unbiased.

However, $\underline{\hat{\mathbf{b}}}$ is a linear estimator itself and thus, $\underline{\tilde{\mathbf{x}}}$ as the sum of linear estimators is a linear estimator defined by Algorithm 2.1 with particular filter and fusion gains. Indeed, filter and fusion gains of $\underline{\tilde{\mathbf{x}}}$ are in general different from the central KF. Therefore, the resulting estimator is suboptimal even if the assumptions of the DKF are satisfied. Hence, instead of applying an additive stochastic bias correction, it is easier to employ the corresponding unbiased local estimator directly.

Indeed, even if it is not viable to correct the bias, the additive technique can still be used to **detect a bias**. To this end, let $\underline{\phi} \in \mathbb{R}^{n_x}$ denote an arbitrary vector known to all nodes in the sensor network. Then, the unknown state $\underline{\mathbf{x}}$ in (4.11), (4.13) and (4.14) can be substituted by $\underline{\phi}$ and an "inspection variable" $\underline{\beta}$ for local estimators $\underline{\hat{\mathbf{x}}}$ is obtained according to

$$\text{Initialization: } \underline{\beta}_0 = \mathrm{E}\{\underline{\hat{\mathbf{x}}}_0 - \underline{\mathbf{x}}_0\} \ , \tag{4.15}$$

$$\text{Prediction: } \underline{\beta}_{k+1} = \mathbf{A}_k \underline{\beta}_k \ , \tag{4.16}$$

$$\text{Filtering: } \underline{\beta}_{k|k} = \mathbf{L}_k \underline{\beta}_k + (\mathbf{L}_k + \mathbf{K}_k \mathbf{H}_k - \mathbf{I}) \underline{\phi} \,, \tag{4.17}$$

$$\text{Fusion: } \underline{\beta}_k = \sum_{s \in \mathcal{S}} \mathbf{F}_k^s \underline{\beta}_k^s + \Big(\sum_{s \in \mathcal{S}} \mathbf{F}_k^s - \mathbf{I} \Big) \underline{\phi} \,. \tag{4.18}$$

Then, $\underline{\beta} = \underline{0}$ indicates that the estimate is unbiased. However, depending on the choice of $\underline{\phi}$, the bias inspection suffers from **false positives**. For example for constant $\underline{\phi} = \underline{1}$, $\underline{\beta}$ indicates a bias-free estimate when the filter operations of two different time steps balance out, even if the fused estimate is biased due to intermediate transformations or differing state variables.

A solution to this problem is to generate time dependent $\underline{\phi}_k$ by means of pseudorandom number generators that are initialized with the same values in all nodes. Then, neglecting quantization effects, false positives are impossible.

On the other hand, a **false negative**, i.e., the inspection variable indicates a bias although the estimate is unbiased, can for example occur when a positive bias offsets a negative bias from a previous time step. However, false negatives are difficult to avoid as this necessitates the quantification and tracking of the true bias, which depends on the unknown true state.

Indeed, if unexpected or unconsidered events cause the fused estimate to be biased, this can be detected by means of the additive debiasing technique.

Example 4.3: Detection of Sensor Failures

Let the scenario be the same as in Example 4.2. The communication error of sensor s_2 is supposed to have been temporary such that after time step 10 the communication works again. For the evaluation an improved DKF is considered that employs the bias detection technique. The estimator at the fusion center predicts the fused estimate from previous time steps as long as a bias is found (DKF+).

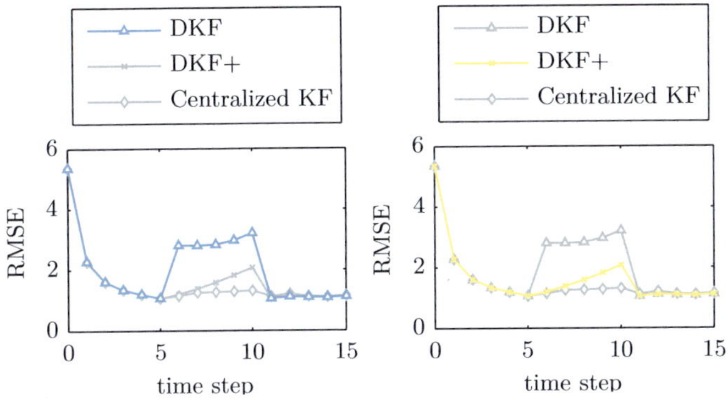

Figure 4.4: RMSE of the fusion center estimate in 100 Monte Carlo runs for DKF, DKF with bias detection handling (DKF+), and centralized KF. In time steps 6 to 10, the fusion center does not receive information from sensor s_2.

The RMSE of the considered estimators is depicted in Figure 4.4. Notably, both versions of the DKF perform better than the centralized KF after time step 10. This is a direct consequence of the preprocessing in distributed estimation. While measurements from time step $6 - 10$ of sensor s_2 are inevitable lost for the calculation of the estimate in the centralized KF, the information is still included in the recursively obtained sensor estimate that is transmitted from sensor s_2 to the fusion center at time step 11 in the DKF processing. In fact, estimates at the fusion center that are provided by the DKF (and DKF+) in time steps 11 to 15 correspond to the centralized KF results that would be obtained when no communication problem had occurred.

The other observation pertains to the difference between DKF and DKF+. As in time steps 6 to 10 no new measurements are filtered when the bias detection technique is utilized, the RMSE increases steadily. However, the error induced by the bias in the DKF is larger, which makes the bias detection extension a meaningful technique to handle rare communication problems. ■

As it has been shown in Example 4.3, an application of the additive bias technique can be justified to handle unexpected events. However, the main drawback of the presented approach manifests itself in the corrective actions. When a bias is detected, the processing at the fusion center cannot use incoming information from working sensors. This is unacceptable for large-scale sensor networks, where it is not only difficult to prevent sensor outages but where data from one sensor has also only little impact on the precision of the fused estimate.

4.1.2 - b Multiplicative Debiasing

As an alternative to the additive bias detection, a **multiplicative bias correction** technique is proposed that permits the incorporation of sensor estimates even if information is received from a subset of sensors only. This theory is used for the estimation in sensor networks with a stochastic communication and with time-varying models in Section 4.2.

In the sequel of this chapter, let the state transition matrix $\mathbf{A}$ be regular. Even though this appears to be a strong assumption, non-invertible state transition matrices imply that at least one variable in the predicted state is redundant. Therefore, another state representation with lower dimension and regular state transition matrix can be found. More details on the reduction of the state dimension are given in [83].

For the multiplicative correction of the bias, a **debiasing matrix** is processed at the sensors. In the interest of simplicity, only the information form of the matrix is considered. However, by means of the

Woodbury matrix inversion lemma [171], the algebraically identical form in the state space can be derived.

Definition 4.2 *Let $\hat{\underline{\mathbf{x}}}$ denote a (potentially) biased estimator. The matrix $\mathbf{\Delta}$ with*

$$\mathrm{E}\{(\mathbf{\Delta})^{-1}\hat{\underline{\mathbf{x}}}\} = \mathrm{E}\{\underline{\mathbf{x}}\} \tag{4.19}$$

is referred to as (multiplicative) ***debiasing matrix*** *of $\hat{\underline{\mathbf{x}}}$.*

Consider a linear estimator from Algorithm 2.1. Then, Definition 4.2 is a sufficient condition to specify the local processing of debiasing matrices. In order to compensate the **initial** transformation $\mathbf{T}\widetilde{\underline{\mathbf{x}}}_0$, the debiasing matrix is set to

$$\mathbf{\Delta}_0 = \mathbf{T}\ . \tag{4.20}$$

Therefore, it holds $\mathrm{E}\{(\mathbf{\Delta}_0)^{-1}\hat{\underline{\mathbf{x}}}_0\} = \mathrm{E}\{(\mathbf{T})^{-1}\mathbf{T}\widetilde{\underline{\mathbf{x}}}_0\} = \mathrm{E}\{\underline{\mathbf{x}}_0\}$ for unbiased estimators $\widetilde{\underline{\mathbf{x}}}_0$. In the **prediction step**, the state evolves according to $\underline{\mathbf{x}}_{k+1} = \mathbf{A}_k\underline{\mathbf{x}}_k + \underline{\mathbf{w}}_k$, where $\mathrm{E}\{\underline{\mathbf{w}}_k\} = \underline{0}$. With the debiasing matrix

$$\mathbf{\Delta}_{k+1} = \mathbf{A}_k\mathbf{\Delta}_k(\mathbf{A}_k)^{-1}\ , \tag{4.21}$$

the correction of time step $k+1$ is traced back to the debiasing from time step k and

$$\begin{aligned}\mathrm{E}\{(\mathbf{\Delta}_{k+1})^{-1}\hat{\underline{\mathbf{x}}}_{k+1}\} &= \mathrm{E}\{\mathbf{A}_k(\mathbf{\Delta}_k)^{-1}(\mathbf{A}_k)^{-1}\mathbf{A}_k\hat{\underline{\mathbf{x}}}_k\}\\ &= \mathbf{A}_k\,\mathrm{E}\{(\mathbf{\Delta}_k)^{-1}\hat{\underline{\mathbf{x}}}_k\}\end{aligned}$$

is satisfied. In the **filter step**, the estimate is transformed according to $\hat{\underline{\mathbf{x}}}_{k|k} = \mathbf{L}_k\hat{\underline{\mathbf{x}}}_k + \mathbf{K}_k\underline{\mathbf{z}}_k$. For

$$\mathbf{\Delta}_{k|k} = \mathbf{L}_k\mathbf{\Delta}_k + \mathbf{K}_k\mathbf{H}_k\ , \tag{4.22}$$

it holds $\mathrm{E}\{(\mathbf{\Delta}_{k|k})^{-1}\hat{\underline{\mathbf{x}}}_{k|k} - \underline{\mathbf{x}}_k\} = \mathrm{E}\{(\mathbf{\Delta}_{k|k})^{-1}(\hat{\underline{\mathbf{x}}}_{k|k} - \mathbf{\Delta}_{k|k}\underline{\mathbf{x}}_k)\}$. Then, unbiasedness follows from the linear estimator processing and the linear measurement model (2.9). It remains to derive the **fusion** of debiasing matrices. For $\hat{\underline{\mathbf{x}}}_k = \sum_{s\in\mathcal{S}} \mathbf{F}_k^s \hat{\underline{\mathbf{x}}}_k^s$, the debiasing matrix

$$\mathbf{\Delta}_k = \sum_{s\in\mathcal{S}} \mathbf{F}_k^s \mathbf{\Delta}_k^s , \tag{4.23}$$

leads to $\mathrm{E}\{\sum_{s\in\mathcal{S}} \mathbf{F}_k^s \hat{\underline{\mathbf{x}}}_k^s\} = \mathrm{E}\{\sum_{s\in\mathcal{S}} \mathbf{F}_k^s \mathbf{\Delta}_k^s \underline{\mathbf{x}}_k\} = \mathrm{E}\{\mathbf{\Delta}_k \underline{\mathbf{x}}_k\}$. In summary, the following statement is obtained.

Theorem 4.2 *Let $\hat{\underline{\mathbf{x}}}_k^s$ denote an arbitrary linear estimator specified by Algorithm 2.1. Then, the linear estimator defined by*

$$\tilde{\underline{\mathbf{x}}}_k^s = (\mathbf{\Delta}_k^s)^{-1} \hat{\underline{\mathbf{x}}}_k^s , \tag{4.24}$$

where the debiasing matrix $\mathbf{\Delta}_k^s$ is obtained according to (4.20) *to* (4.23) *is unbiased.*

The special case $\mathbf{\Delta} = \mathbf{I}$ implies $\mathrm{E}\{\hat{\underline{\mathbf{x}}}\} = \mathrm{E}\{\underline{\mathbf{x}}_k\}$, i.e., the unbiasedness of the estimator $\hat{\underline{\mathbf{x}}}$. Hence, a bias detection as it has been considered in the last section can also be realized with a multiplicative bias correction technique. This insight is captured in the following corollary.

Corollary 4.3 *The estimator $\hat{\underline{\mathbf{x}}}$ from Theorem 4.2 is unbiased, if and only if $\mathbf{\Delta} = \mathbf{I}$.*

Moreover, the **information form** of the debiasing matrix allows for the processing of underdetermined pseudo-estimates that emerge, e.g., when no initial estimates are provided. For this purpose, it is sufficient to set $\mathbf{\Delta}_0 = \mathbf{0}$. Then, the actual initialization is carried out by means of the first measurement and a reconstruction of the actual state estimate is possible when $\mathbf{\Delta}$ becomes invertible in subsequent operations.

In summary, the processing of one debiasing matrix per sensor suffices to trace and remove the bias of an estimate in sensor networks. This property is exploited in the next sections to derive efficient estimators for distributed systems.

4.2 The Hypothesizing Distributed Kalman Filter

In this section, the DKF is generalized by means of the multiplicative debiasing technique. As a result, the constraints concerning global model knowledge are relaxed. More precisely, the **hypothesizing distributed Kalman filter** is derived, which provides globally optimized filter gains subject to limited model knowledge, stochastic communication, and sensor failures.

The next example illustrates a typical surveillance and tracking scenario realized with a sensor network and serves as a motivation for the subsequent considerations.

Example 4.4: Target Tracking in 2D with Local KFs

The challenge is to track the position and velocity of a target in a two-dimensional area. Let the target's initial state be $(50, 3, 50, 3)$ (x pos., x vel., y pos., y vel.) and assume that the state's dynamic is specified by a disturbed rotational motion model [183], i.e.,

$$\underline{\mathbf{x}}_{k+1} = \begin{pmatrix} 1 & \frac{\sin(\omega)}{\omega} & 0 & \frac{-1+\cos(\omega)}{\omega} \\ 0 & \cos(\omega) & 0 & -\sin(\omega) \\ 0 & \frac{1-\cos(\omega)}{\omega} & 1 & \frac{\sin(\omega)}{\omega} \\ 0 & \sin(\omega) & 0 & \cos(\omega) \end{pmatrix} \underline{\mathbf{x}}_k + \underline{\mathbf{w}}_k \ ,$$

where $\underline{\mathbf{w}}_k \sim \mathcal{N}(\underline{0}, \mathbf{Q})$, and

$$\mathbf{Q} = \begin{pmatrix} 0.8 & 0 & 0.1 & 0.2 \\ 0 & 0.3 & 0 & 0 \\ 0.1 & 0 & 0.7 & 0 \\ 0.2 & 0 & 0 & 0.4 \end{pmatrix} ,$$

with $\omega = 0.1$. The position of the target is observed by 40 sensors that are randomly placed in the area $[0, 100] \times [0, 100]$.

Sensors obtain measurements according to the distance-dependent model

$$\underline{\mathbf{z}}_k^s = \begin{pmatrix} 1 & 0 & 0 & 0 \\ 0 & 0 & 1 & 0 \end{pmatrix} \underline{\mathbf{x}}_k + \underline{\mathbf{v}}_k \text{ with } \underline{\mathbf{v}}_k \sim \mathcal{N}(\underline{0}, \mathbf{R}(\underline{\mathbf{x}}_k)) ,$$

where

$$\mathbf{R}(\underline{\mathbf{x}}_k) = \ell_2 \begin{pmatrix} \underline{\mathbf{x}}_k(1) - p_x^s \\ \underline{\mathbf{x}}_k(3) - p_y^s \end{pmatrix} \cdot \begin{pmatrix} 5 & 2 \\ 2 & 5 \end{pmatrix} ,$$

and where p_x^s and p_y^s denote x and y position of sensor s. The sensors are initialized with independent estimates specified by $\mathbf{C}_0^s = 20 \cdot \mathbf{I}$ and track the target with local KFs over 50 time steps. At each time step, estimates from all sensors are transmitted to the fusion center, which fuses the received data by means of the LMMSE combination from Theorem 2.5.

Note that under realistic conditions, cross-covariance matrices are unknown to the fusion center so that the LMMSE fusion method would not be applicable (c.f., Chapter 3). However, in order to focus on filter gains, an omniscient observer is assumed to realize the baseline for local KF approaches. The theoretic baseline for linear estimators with an arbitrary local processing is obtained by means of a centralized KF.

The results of one run are illustrated in Figure 4.5. For the evaluation of the RMSE, the covariance of the estimate is calculated by means of the joint space representation from Section 2.3.4 in closed form.

Three observations are striking. First, the RMSE and the measurement capacity $\mathbf{C}^z$ from (4.9) are (strongly) negatively correlated. In particular, the measurement capacity increases in areas with dense coverage, while the RMSE declines in such regions.

Second, the local measurement models vary with the distance to the target, but the measurement capacity as the sum of all these local matrices remains almost constant.

Third, filter gains for the local KF processing are calculated with measurement capacities that are smaller than or equal to the ones depicted in the top right part of Figure 4.5. This stands in contrast to the centralized KF, which utilizes the true measurement capacity and yields significantly lower estimation errors. ■

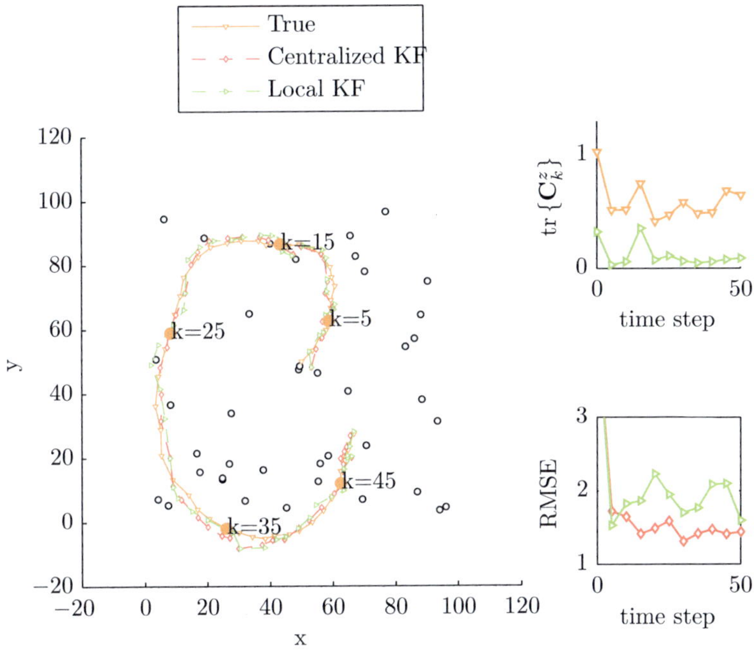

Figure 4.5: The results of one run of the tracking simulation from Example 4.4. In the left figure, the true and the estimated paths are depicted. Sensor positions are given as grey points. In the top right figure, the trace of the true measurement capacity $\mathrm{tr}\left\{\sum_{s\in\mathcal{S}}(\mathbf{H}_k^s)^\top(\mathbf{R}_k^s)^{-1}\mathbf{H}_k^s\right\}$ and the trace of the maximum local measurement capacity $\max_{s\in\mathcal{S}}\mathrm{tr}\left\{(\mathbf{H}_k^s)^\top(\mathbf{R}_k^s)^{-1}\mathbf{H}_k^s\right\}$ are compared. In the bottom right figure, the RMSE at the fusion center is depicted for all time steps.

A key observation from Example 4.4 is the **permanence of the global measurement capacity**. In fact, quite often this permanence is a desirable design attribute of sensor networks. In order to monitor an area with a guaranteed quality subject to minimal costs, sensors must be deployed in a way that the global measurement capacity is similar at all locations.

For the derivation of a distributed estimator that accomplishes high precision subject to imprecise model information, the DKF is generalized. As a matter of fact, the LMMSE estimator is achieved when distributed sensors utilize filter gains with the correct matrix-relation between each other. Therefore, the objective must be to imitate centralized KF gains at the distributed nodes.

Now, the idea is to generalize the DKF by treating the measurement capacity $\mathbf{C}_k^z$ as an application-specific matrix parameter, which can be optimized. This is feasible since the measurement capacity in large sensor networks is non-erratic, as it has been motivated in Example 4.4. Then, in general, bias terms emerge due to discrepancies between hypothesis and true measurement capacities that are traced and corrected with the multiplicative debiasing technique from Section 4.1.2. Overall, the following main topics are captured.

- The **hypothesizing distributed Kalman filter** is proposed as generalization of the DKF.
- Fusion techniques are discussed when **packet losses** occur during communication.
- The sensor processing is generalized to delay-dependent measurement capacities in order to account for **packet delays**.
- A **feedback technique** for the optimization of the hypothesis based on debiasing matrices is derived.

The resulting estimator optimizes filter gains at the sensors by means of a hypothesis about the (global) measurement capacity. By this means, LMMSE estimates are achieved for a properly chosen hypothesis. In this section, the estimator processing is examined when all sensors apply the same hypothesis. A more general perspective on the optimization of filter gains with differing hypotheses is given in the next section.

4.2.1 Motivation and Basic Processing

The **hypothesizing distributed Kalman filter** (HKF) consists of a sensor processing and a technique for the combination of estimates at the fusion center. The algorithm and the associated theory have been developed in [175,182–185,187]. For the application of the algorithm, a parameter $\hat{\mathbf{C}}^z$ is employed to calculate a hypothesized global covariance according to which filter gains can be optimized. In order to achieve LMMSE estimates at the fusion center, this parameter is supposed to satisfy

$$\hat{\mathbf{C}}^z \stackrel{!}{=} \mathbf{C}_k^z = \sum_{s \in \mathcal{S}} (\mathbf{H}_k^s)^\top (\mathbf{R}_k^s)^{-1} \mathbf{H}_k^s$$

and must be known to all sensors.

4.2.1 - a Sensor Processing

Let initial estimates $\underline{\widetilde{\mathbf{x}}}_0^s$ be independent with covariance $\mathbf{C}_0^s$. A **hypothesized global covariance** is recursively obtained at the sensors according to

$$\hat{\mathbf{C}}_0 = \Big(\sum_{s \in \mathcal{S}} (\mathbf{C}_0^s)^{-1} \Big)^{-1} , \tag{4.25}$$

$$\hat{\mathbf{C}}_{k+1} = \mathbf{A}_k \hat{\mathbf{C}}_k (\mathbf{A}_k)^\top + \mathbf{Q}_k , \tag{4.26}$$

$$\hat{\mathbf{C}}_{k|k} = ((\hat{\mathbf{C}}_k)^{-1} + \hat{\mathbf{C}}^z)^{-1} , \tag{4.27}$$

where $\hat{\mathbf{C}}^z$ denotes the hypothesis about the measurement capacity. As the covariance $\hat{\mathbf{C}}$ usually differs from the true global covariance of the DKF, it is denoted with a hat. In fact, with (4.25) to (4.27) the covariance of a centralized KF with $\mathbf{C}_k^z = \hat{\mathbf{C}}^z$, $k \in \mathcal{K}$ is calculated. By means of the estimated covariance, (KF) gains are determined according to

$$\mathbf{L}_k = \hat{\mathbf{C}}_{k|k}(\hat{\mathbf{C}}_k)^{-1} , \tag{4.28}$$

$$\mathbf{K}_k^s = \hat{\mathbf{C}}_{k|k}(\mathbf{H}_k^s)^\top(\mathbf{R}_k^s)^{-1} . \tag{4.29}$$

Then, the processing of estimates follows Algorithm 2.1 with gains (4.28) and (4.29) and $\mathbf{T}^s = \hat{\mathbf{C}}_0(\mathbf{C}_0^s)^{-1}$. Note that in contrast to the DKF, gains are obtained by means of the hypothesis $\hat{\mathbf{C}}^z$ and are independent from measurement models of remote nodes. The calculation of the debiasing matrix is specified by the filter gains and follows along the lines of (4.20) to (4.22). The overall processing is summarized in Algorithm 4.4.

Algorithm 4.4 Sensor Processing of the HKF

1: Initialization: $(\hat{\underline{\mathbf{x}}}_0, \hat{\mathbf{C}}_0, \mathbf{\Delta}_0) = (\hat{\mathbf{C}}_0(\mathbf{C}_0)^{-1}\tilde{\underline{\mathbf{x}}}_0, \hat{\mathbf{C}}_0, \hat{\mathbf{C}}_0(\mathbf{C}_0)^{-1})$ where $\hat{\mathbf{C}}_0 = \left(\sum_{s\in\mathcal{S}}(\mathbf{C}_0^s)^{-1}\right)^{-1}$
2: **for** $k = 1; k \in \mathcal{K}; k = k+1$ **do**
3: Prediction: $(\hat{\underline{\mathbf{x}}}_k, \hat{\mathbf{C}}_k, \mathbf{\Delta}_k) = (\mathbf{A}_{k-1}\hat{\underline{\mathbf{x}}}_{k-1}, \mathbf{A}_{k-1}\hat{\mathbf{C}}_{k-1}(\mathbf{A}_{k-1})^\top + \mathbf{Q}_{k-1}, \mathbf{A}_{k-1}\mathbf{\Delta}_{k-1}(\mathbf{A}_{k-1})^{-1})$
4: Filtering: $(\hat{\underline{\mathbf{x}}}_{k|k}, \hat{\mathbf{C}}_{k|k}, \mathbf{\Delta}_{k|k}) = (\mathbf{L}_k\hat{\underline{\mathbf{x}}}_k + \mathbf{K}_k\underline{\mathbf{z}}_k, ((\hat{\mathbf{C}}_k)^{-1} + \hat{\mathbf{C}}^z)^{-1}, \mathbf{L}_k\mathbf{\Delta}_k + \mathbf{K}_k\mathbf{H}_k)$ with $\mathbf{L}_k, \mathbf{K}_k$ from (4.28) and (4.29)
5: **end for**

Considering Theorem 4.2, sensors obtain unbiased local estimates according to $\tilde{\underline{\mathbf{x}}} = (\mathbf{\Delta})^{-1}\hat{\underline{\mathbf{x}}}$. Multiple measurements in one time step or missing measurements can be handled by adding or omitting $\mathbf{K}_k\underline{\mathbf{z}}_k$ terms in the estimate and $\mathbf{K}_k\mathbf{H}_k$ terms in the debiasing matrix equations.

4.2.1 - b Fusion Center Processing

As in the DKF, the fusion is realized by means of a replacement policy. For a reconstruction of the estimate at the fusion center, estimates and debiasing matrices must be exchanged. The fusion equation is adopted from the DKF and is given by

$$\underline{\hat{\mathbf{x}}}_k = \sum_{s \in \mathcal{S}} \underline{\hat{\mathbf{x}}}_k^s \, . \tag{4.30}$$

In terms of Algorithm 2.1, the summation in (4.30) corresponds to fusion weights $\mathbf{F}^s{=}\mathbf{I}$ so that the debiasing matrix follows with (4.23) as

$$\mathbf{\Delta}_k = \sum_{s \in \mathcal{S}} \mathbf{\Delta}_k^s \, . \tag{4.31}$$

An unbiased estimate at the fusion center is reconstructed according to $\hat{\underline{\mathbf{x}}}_k = (\mathbf{\Delta}_k)^{-1}\underline{\hat{\mathbf{x}}}_k$. As the combination of debiasing matrices (4.31) is **additive** (4.30), the in-network processing in hierarchical sensor networks from Section 4.1.1 is possible.

When the hypothesis $\hat{\mathbf{C}}^z$ equals the true measurement capacity, the HKF corresponds to the DKF. Note that in this case, the debiasing matrix vanishes according to Corollary 4.3. This leads to the following corollary of Theorem 4.1.

Corollary 4.4 *Let the sensors apply Algorithm 4.4 and let* $\mathbf{C}_k^z = \hat{\mathbf{C}}^z$, $k \in \mathcal{K}$. *Then,* $\hat{\underline{\mathbf{x}}} = \sum_{s \in \mathcal{S}} \underline{\hat{\mathbf{x}}}^s$ *is the LMMSE estimator.*

4.2.2 Packet Delays and Losses

Packet losses caused by concurrent transmissions, environmental influences, or other disturbances are common phenomena in sensor networks. As discussed in Section 2.3.2, one of the key drivers for distributed estimation is its capability to cushion the impact of packet losses. While information is inevitably lost when the transmission of a measurement in a UDP-like[2] network fails, local estimates at the sensors comprise all information observed up to the latest time step. Therefore, a subsequent transmission of the sensor estimate of a later time step still entails past measurement information.

Apart from packet losses, communication time in large sensor networks is often not negligible. Again, there are myriad reasons, ranging from actual transmission time on limited frequencies over blocking times induced by communication protocols. In particular, in hop-to-hop networks that involve the routing of packets, transmission of data can require a considerable amount of time [21].

4.2.2 - a Packet Losses

By means of the implementation of the HKF from Section 4.2.1, rare sensor failures can be handled. Indeed, due to the replacement policy, packets received prior to the latest time step are ignored in the calculation of the fused estimate. This is an appropriate policy when almost no packet delays or losses occur as then all information from sensor s is stored in the latest estimate anyway. However, when the latest estimates from several sensors are missing, past estimates from time steps $\tilde{k} < k$ carry a significant amount of information about the true state and should be used in the calculation of the fused estimate.

An efficient technique that exploits past estimates is to store sensor-specific **proxy estimates** $\breve{\underline{\mathbf{x}}}^s$ with debiasing matrices $\breve{\mathbf{\Delta}}^s$ at the fusion center. When an estimate $\hat{\underline{\mathbf{x}}}^s$ is received from sensor s, the proxy estimate $\breve{\underline{\mathbf{x}}}^s$ is replaced. Otherwise, the proxy estimate is processed like a sensor estimate that is not updated with measurements. As discussed in Section 4.2.1, the corresponding filter processing is given by

$$\breve{\underline{\mathbf{x}}}^s_{k|k} = \mathbf{L}_k \breve{\underline{\mathbf{x}}}^s_k \text{ and } \breve{\mathbf{\Delta}}^s_{k|k} = \mathbf{L}_k \breve{\mathbf{\Delta}}^s_k \ .$$

[2]User Datagram Protocol [132]: a popular protocol without a handshake mechanism, i.e., without confirmation and re-transmission of data.

Then, proxy estimates can be used as **substitutes for estimates** that are not received in the latest time step. However, they only correspond to sensor estimates when no packet losses occurred and otherwise lack the measurement information from all time steps since the last successful transmission. The proposed technique is summarized in Algorithm 4.5.

Algorithm 4.5 Fusion Center Processing of the HKF

```
Initialize: $\mathcal{X} = \{(\breve{\underline{\mathbf{x}}}_0^s, \breve{\mathbf{\Delta}}_0^s)\}_{s\in\mathcal{S}}$, $\hat{\mathbf{C}}_0 = \left(\sum_{s\in\mathcal{S}}(\mathbf{C}_0^s)^{-1}\right)^{-1}$
for $k = 1; k \in \mathcal{K}; k = k+1$ do
    $\mathbf{L}_k = \hat{\mathbf{C}}_{k|k}(\hat{\mathbf{C}}_{k-1})^{-1}$ with $\hat{\mathbf{C}}_k$ from (4.26) and (4.27)
    for $s \in \mathcal{S}$ do
        if Packet $\mathcal{P}_k^s = (\hat{\underline{\mathbf{x}}}_k^s, \mathbf{\Delta}_k^s)$ is received then
            $\{\mathcal{X}\}_s = \mathcal{P}_k^s$
        else
            $\{\mathcal{X}\}_s = (\mathbf{L}_k\mathbf{A}_{k-1}\breve{\underline{\mathbf{x}}}_k^s, \mathbf{L}_k\mathbf{A}_{k-1}\breve{\mathbf{\Delta}}_k^s(\mathbf{A}_{k-1})^{-1})$
        end if
    end for

    $(\hat{\underline{\mathbf{x}}}_k, \mathbf{\Delta}_k) = (\sum_{s\in\mathcal{S}}\breve{\underline{\mathbf{x}}}_k^s, \sum_{s\in\mathcal{S}}\breve{\mathbf{\Delta}}_k^s)$
    $\widetilde{\underline{\mathbf{x}}}_k = (\mathbf{\Delta}_k)^{-1}\hat{\underline{\mathbf{x}}}_k$
end for
```

The hypothesis in a sensor network that is subject to packet losses should be calculated with regard to the stochastic properties of the packet arrival process. Let $\underline{\mu}_k^s$ denote an indicator variable that specifies whether the transmission of sensor s to the fusion center in time step k is successful. Then, the hypothesis can be obtained according to the expected measurement capacity

$$\hat{\mathbf{C}}^z \stackrel{!}{=} \mathrm{E}\left\{\sum_{s\in\mathcal{S}} \underline{\mu}_k^s (\mathbf{H}_k^s)^\top (\mathbf{R}_k^s)^{-1}\mathbf{H}_k^s\right\} .$$

Note that this approach is only realizable when stochastic properties of the communication network are known. A more detailed discussion on the choice of the hypothesis is given in Section 4.2.3. Evaluations of the HKF with imperfectly chosen hypotheses, e.g., in the presence of false measurements and with the Ornstein-Uhlenbeck model, are presented in [29, 35] and by means of the next example.

Example 4.5: Target Tracking in 2D with the HKF

For the evaluation of the HKF in stochastic communication networks, the two-dimensional tracking scenario from Example 4.4 is extended. Instead of assuming a stable connection between sensors and fusion center, an energy-costly wireless communication is considered. More precisely, the sensors transmit estimates only each 5th time step beginning in the 5th time step and the wireless communication causes a packet loss rate of 20%.

HKF and local KF processing in combination with the LMMSE combination from Theorem 2.5 are considered. As discussed in Example 4.4, the LMMSE fusion is not applicable and is merely used to provide a bound on the precision of the local KF processing. In contrast to that, the HKF utilizes a suboptimal fusion formula that can be calculated at the fusion center without cross-covariance matrices.

For the calculation of a suitable hypothesis, it is possible to examine the expected measurement capacity at each position of the observed area and determine the average under consideration of the packet dropout rate. However, for the sake of simplicity, the value is heuristically observed to be

$$\hat{\mathbf{C}}^z = \frac{1}{35}(\mathbf{H}^s)^\top \begin{pmatrix} 5 & 2 \\ 2 & 5 \end{pmatrix}^{-1} \mathbf{H}^s \, .$$

The tracking results averaged over 100 Monte Carlo runs are given in Figure 4.6. As illustrated in the top right part of Figure 4.6, the hypothesis matches the measurement capacity well. Thus, the HKF estimate is close to the LMMSE estimate. Especially, if many sensors

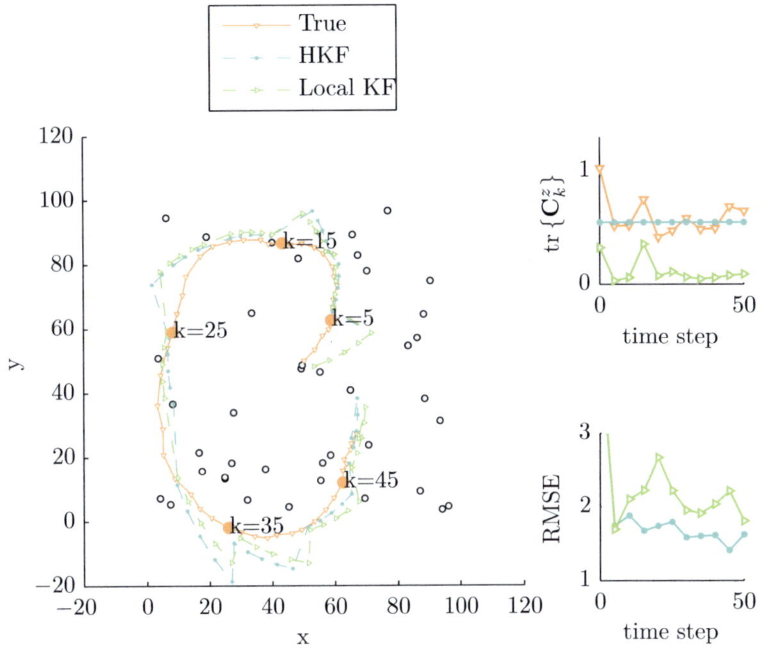

Figure 4.6: The results of an extended version of the scenario from Example 4.4. As illustrated in the top right figure, the true measurement capacity of the sensor network is approximately met by the hypothesis. Consequently, the RMSE of the HKF is considerably lower than for the local KF processing.

contribute information about the state, i.e., if the target is close to a high number of sensors, as for example in time step 45, the HKF distinctively outperforms the local KF processing.

Note that the presented results for the local KF processing define a lower bound on the achievable performance while real-world techniques rely on estimated covariances and provide a higher MSE. ■

4.2.2 - b Packet Delays

In the following, the communication delay of estimates is explicitly considered. For this purpose, let sensors be grouped according to the communication time to the fusion center as depicted in Figure 4.7. Such a model covers, in particular, hop-to-hop networks where the transmission of data requires one time step.

A **naïve technique** to compensate for the effect of packet delays is to wait for estimates of the same time step at the fusion center. Indeed, this is suboptimal as no estimate is provided for the latest time step and smoothing techniques yield better results by incorporating all packets received in the waiting time.

The HKF extension for the handling of stochastic packet delays is derived successively. Beginning with a central filter algorithm that provides LMMSE estimates for deterministic communication networks, the calculations are distributed, and stochastic packet delays are introduced. The algorithms were proposed in [187] and are presented here in a short version.

Centralized Kalman filter In a first step, a centralized KF that receives delayed measurements is considered. Let the upper bound on the packet delay be denoted by $\bar{\tau}$, and let the packet delays $\tau \in \{0, \dots, \bar{\tau} - 1\}$ be fixed for each sensor[3]. Then, at time step k, measurements from time steps $[k + 1 - \bar{\tau}, k]$ are received.

The processing of those out-of-sequence measurements has already been discussed in Section 2.3.1. In particular, it is possible to reconstruct the LMMSE estimate of time step k with a centralized KF by recursively comprising measurements up to time step $k - \bar{\tau}$ and by applying temporary gains to measurements from time steps $k - \bar{\tau} + 1$ to k.

[3]For the general applicability of the approach, $\tau = 0$ is considered as well.

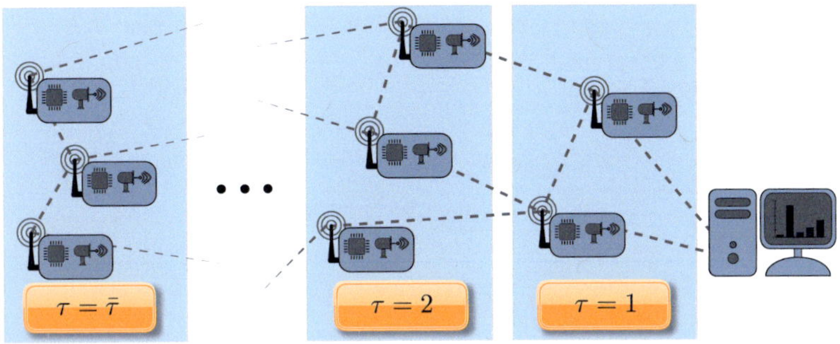

Figure 4.7: A schematic overview of a hop-to-hop network with sensors and fusion center. When the communication is deterministic, the nodes can be grouped according to the minimal communication time to the fusion center.

Let k be fixed and let $\hat{\underline{\mathbf{x}}}_{k-\bar{\tau}}$ denote the estimate of a centralized KF with covariance $\mathbf{C}_{k-\bar{\tau}}$. When only packet delays are considered and the sensors are grouped as depicted in Figure 4.7, measurements of time step k are only available from those sensors that are in the group $\tau = 0$, measurements of time step $k-1$ are available of sensors from the groups $\tau = 0$ and $\tau = 1$, and so forth. For the centralized KF, the corresponding covariance can be obtained recursively beginning at time step $\tilde{k} = k - \bar{\tau}$ with

$$\breve{\mathbf{C}}_{\tilde{k}+1} = \mathbf{A}_{\tilde{k}} \breve{\mathbf{C}}_{\tilde{k}} (\mathbf{A}_{\tilde{k}})^\top + \mathbf{Q}_{\tilde{k}} \,, \tag{4.32}$$

$$\breve{\mathbf{C}}_{\tilde{k}|\tilde{k}} = \Big((\breve{\mathbf{C}}_{\tilde{k}})^{-1} + \sum_{s \in \mathcal{S}^{k-\tilde{k}}} (\mathbf{H}^s_{\tilde{k}})^\top (\mathbf{R}^s_{\tilde{k}})^{-1} \mathbf{H}^s_{\tilde{k}} \Big)^{-1} \,, \tag{4.33}$$

where $\mathcal{S}^\tau \subseteq \mathcal{S}$ denotes the set of sensors that transmit measurements with a delay of equal or less than τ to the fusion center. According to (4.1), the filter gains of the centralized KF are given by

$$\breve{\mathbf{L}}_{\tilde{k}} = \breve{\mathbf{C}}_{\tilde{k}|\tilde{k}} (\breve{\mathbf{C}}_{\tilde{k}})^{-1} \text{ and } \breve{\mathbf{K}}^s_{\tilde{k}} = \breve{\mathbf{C}}_{\tilde{k}|\tilde{k}} (\mathbf{H}^s_{\tilde{k}})^\top (\mathbf{R}^s_{\tilde{k}})^{-1} \,.$$

Therefore, the measurement processing of the centralized KF with an LMMSE out-of-sequence handling is recursively specified based on $\breve{\underline{\mathbf{x}}}_{k-\bar{\tau}} = \hat{\underline{\mathbf{x}}}_{k-\bar{\tau}}$ by

$$\begin{aligned}\breve{\underline{\mathbf{x}}}_{\tilde{k}+1} &= \mathbf{A}_{\tilde{k}} \breve{\underline{\mathbf{x}}}_{\tilde{k}} \ , \\ \breve{\underline{\mathbf{x}}}_{\tilde{k}|\tilde{k}} &= \breve{\mathbf{L}}_{\tilde{k}} \breve{\underline{\mathbf{x}}}_{\tilde{k}} + \sum_{s \in \mathcal{S}^{k-\tilde{k}}} \breve{\mathbf{K}}^s_{\tilde{k}} \underline{\mathbf{z}}^s_{\tilde{k}} \ ,\end{aligned}$$

for $\tilde{k} \in \{k - \bar{\tau}, k\}$.

Distributed Kalman filter In the next step, measurements comprised in the estimate of the centralized KF are clustered according to sensor sources. To this end, let sensor s be subject to a delay of τ and assume that measurements of the time period $[0, \ldots, k - \bar{\tau}]$ are subsumed in the variable $\hat{\underline{\mathbf{x}}}^s_{k-\bar{\tau}}$. Then, the contribution of sensor s to the fused estimate $\breve{\underline{\mathbf{x}}}_k$ is recursively obtained with $\breve{\underline{\mathbf{x}}}^s_{k-\bar{\tau}} = \hat{\underline{\mathbf{x}}}^s_{k-\bar{\tau}}$ according to

$$\breve{\underline{\mathbf{x}}}^s_{\tilde{k}+1} = \mathbf{A}_{\tilde{k}} \breve{\underline{\mathbf{x}}}^s_{\tilde{k}} \ , \tag{4.34}$$

$$\breve{\underline{\mathbf{x}}}^s_{\tilde{k}|\tilde{k}} = \breve{\mathbf{L}}_{\tilde{k}} \breve{\underline{\mathbf{x}}}^s_{\tilde{k}} + \delta_{\tilde{k} \leq k-\tau} \breve{\mathbf{K}}^s_{\tilde{k}} \underline{\mathbf{z}}^s_{\tilde{k}} \ , \tag{4.35}$$

where $\delta_{\tilde{k} \leq k-\tau}$ is 1 for $\tilde{k} \leq k - \tau$ and 0 otherwise. Suppose that for each sensor s, a vector $\breve{\underline{\mathbf{x}}}^s_k$ is obtained according to the recursive processing from (4.34) and (4.35). Then, it is easy to verify that $\breve{\underline{\mathbf{x}}}_k = \sum_{s \in \mathcal{S}} \breve{\underline{\mathbf{x}}}^s_k$ holds. Therefore, for fixed k, the calculation of the estimate of the centralized KF with delayed measurements can be decomposed into sensor-specific vectors $\breve{\underline{\mathbf{x}}}^s_k$.

Now, consider the transition from time step k to $k + 1$. The covariances, gains, and estimates in the formulas above have already been marked with a breve to indicate the **temporary character**. As for $\tau < \bar{\tau}$, the groups $\mathcal{S}^{k-\tilde{k}}$ grow in number with advancing k for fixed $\tilde{k}$, the matrices and vectors implicitly depend on k and must be recalculated at every time step.

However, as $\mathcal{S}^{\bar{\tau}} = \mathcal{S}$, no additional measurements are received for time step $k+1-\bar{\tau}$ in subsequent processing steps. Therefore, the variables $\hat{\underline{\mathbf{x}}}^s_{k-\bar{\tau}}$ and $\mathbf{C}_{k-\bar{\tau}}$ can be predicted and filtered conclusively to time step $k+1-\bar{\tau}$. It follows that the same considerations permit the conclusive prediction and filtering of the estimates $\hat{\underline{\mathbf{x}}}^s_{k-\bar{\tau}}$. The temporary estimates $\breve{\underline{\mathbf{x}}}^s$ are then processed based on $\hat{\underline{\mathbf{x}}}^s_{k+1-\bar{\tau}}$ and $\mathbf{C}_{k+1-\bar{\tau}}$ with (4.34) and (4.35) up to time step $k+1$ such that the estimate $\breve{\underline{\mathbf{x}}}_{k+1}$ is obtained as $\sum_{s\in\mathcal{S}} \breve{\underline{\mathbf{x}}}^s_{k+1}$.

The derived processing can be defined in a **nested loop** within the DKF. Let $k^s = k-\tau$ denote the time step at which sensor s processes the packet that is received by the fusion center at time step k. In the outer loop, the sensors process the vectors $\hat{\underline{\mathbf{x}}}^s_{\tilde{k}}$ and covariances $\mathbf{C}_{\tilde{k}}$ recursively up to time step $k-\bar{\tau}$ with the (basic) DKF and store the measurements of subsequent time steps $\tilde{k} \in \{k+1-\bar{\tau}, \ldots, k^s\}$ in a set. Then, at time step k^s, $\hat{\underline{\mathbf{x}}}^s_{k-\bar{\tau}}$, $\mathbf{C}_{k-\bar{\tau}}$, and measurements $\underline{\mathbf{z}}^s_{k+1-\bar{\tau}}, \ldots, \underline{\mathbf{z}}^s_{k^s}$ are available at sensor s.

Based on those quantities, in each time step, temporary variables $\breve{\mathbf{C}}_{k-\bar{\tau}} = \mathbf{C}_{k-\bar{\tau}}$, $\breve{\underline{\mathbf{x}}}^s_{k-\bar{\tau}} = \hat{\underline{\mathbf{x}}}^s_{k-\bar{\tau}}$ are initialized and processed with (4.32) to (4.35) up to time step k. Finally, $\breve{\underline{\mathbf{x}}}^s_k$ is transmitted to the fusion center at time step k^s and all temporary variables are discarded. The processing is depicted in Figure 4.8.

At the **fusion center**, the received packages are summed up according to $\breve{\underline{\mathbf{x}}}_k = \sum_{s\in\mathcal{S}} \breve{\underline{\mathbf{x}}}^s_k$ such that the fused estimate equals the centralized KF estimate. The communication effort is the same as for the basic version of the DKF as only the vectors $\breve{\underline{\mathbf{x}}}^s_k$ need to be transmitted from the sensors to the fusion center.

However, note that each sensor has to process a temporary estimate and covariance for all time steps $\tilde{k} \in \{k+1-\bar{\tau}, \ldots, k\}$ such that the computational load can be considerable when the maximum transmission delay $\bar{\tau}$ is large. Therefore, techniques to minimize the computational load by outsourcing calculations to the fusion center or deriving steady state values are discussed in [187].

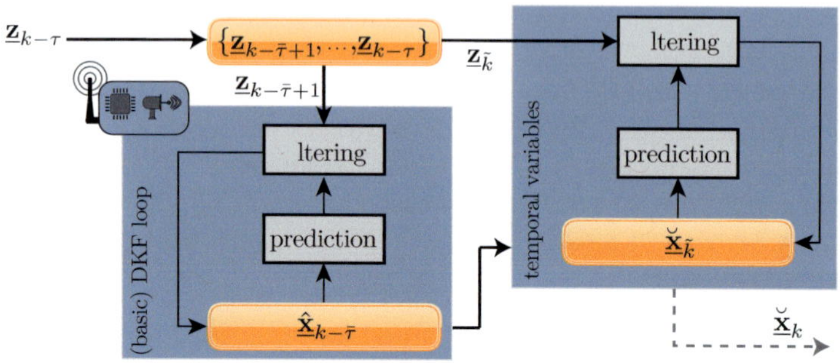

Figure 4.8: The two calculation loops for the delayed DKF. On the left side, the (basic) DKF recursively obtains an estimate with delay $\bar{\tau}$ in one iteration per time step. On the right side, the calculation of the temporary quantity $\breve{\underline{\mathbf{x}}}_{\tilde{k}}$ is illustrated. In each time step, the variable $\breve{\underline{\mathbf{x}}}_{k-\bar{\tau}}$ is filtered and processed to time step k.

By means of the delayed DKF, the calculation of the LMMSE estimate is distributed. In particular, in the presence of rare packet losses or when estimates are required at a lower frequency than measurements are observed, the distributed calculation improves the precision and lowers the communication costs compared to measurement transmission techniques significantly. The same advantages that apply to the DKF also apply to the generalization that takes into account packet delays. However, optimality could be achieved only based on global model knowledge.

Hypothesizing Distributed Kalman filter The filter gain calculation of the HKF without packet delays relies on a hypothesized measurement capacity. According to (4.32) and (4.33), the covariance of the centralized KF with packet delays is obtained in the filter step according to

$$\breve{\mathbf{C}}_{\tilde{k}|\tilde{k}} = \left((\breve{\mathbf{C}}_{\tilde{k}})^{-1} + \mathbf{C}^{z}_{k-\tilde{k}}\right)^{-1} ,$$

where the measurement capacity $\mathbf{C}_{\tau}^{z}$ with delay $\tau \in \{0, \ldots, \bar{\tau}-1\}$ is given by

$$\mathbf{C}_{k-\tilde{k}}^{z} = \sum_{s \in \mathcal{S}^{k-\tilde{k}}} (\mathbf{H}_{\tilde{k}}^{s})^{\top} (\mathbf{R}_{\tilde{k}}^{s})^{-1} \mathbf{H}_{\tilde{k}}^{s} \ .$$

In fact, when packet losses are neglected and time-invariant systems are considered, it holds $\mathbf{C}_{0}^{z} \leq \mathbf{C}_{1}^{z} \leq \cdots \leq \mathbf{C}_{\bar{\tau}-1}^{z}$. Evidently, assuming the same hypothesis for all delays is suboptimal so that the HKF for packet delays requires multiple hypotheses with $\hat{\mathbf{C}}_{0}^{z} \overset{!}{=} \mathbf{C}_{0}^{z}, \ldots, \hat{\mathbf{C}}_{\bar{\tau}-1}^{z} \overset{!}{=} \mathbf{C}_{\bar{\tau}-1}^{z}$.

Let the hypotheses $\hat{\mathbf{C}}_{0}^{z}, \ldots, \hat{\mathbf{C}}_{\bar{\tau}-1}^{z}$ be available to all sensor nodes. Then, the processing of sensor estimates is identical to the DKF with packet delays, which is depicted in Figure 4.8, except that the true measurement capacities in (4.33) are replaced with the corresponding hypotheses, i.e., the covariance is obtained with

$$\tilde{\mathbf{C}}_{\tilde{k}+1} = \mathbf{A}_{\tilde{k}} \tilde{\mathbf{C}}_{\tilde{k}} (\mathbf{A}_{\tilde{k}})^{\top} + \mathbf{Q}_{\tilde{k}} \ , \tag{4.36}$$

$$\tilde{\mathbf{C}}_{\tilde{k}|\tilde{k}} = \left((\tilde{\mathbf{C}}_{\tilde{k}})^{-1} + \hat{\mathbf{C}}_{k-\tilde{k}}^{z} \right)^{-1} \ , \tag{4.37}$$

and filter gains with

$$\tilde{\mathbf{L}}_{\tilde{k}} = \tilde{\mathbf{C}}_{\tilde{k}|\tilde{k}} (\tilde{\mathbf{C}}_{\tilde{k}})^{-1} \text{ and } \tilde{\mathbf{K}}_{\tilde{k}}^{s} = \tilde{\mathbf{C}}_{\tilde{k}|\tilde{k}} (\mathbf{H}_{\tilde{k}}^{s})^{\top} (\mathbf{R}_{\tilde{k}}^{s})^{-1} \ ,$$

where variables with $\tilde{k} > k - \bar{\tau}$ are temporary. The properties of the basic HKF transfer to the delayed version naturally so that for $\hat{\mathbf{C}}_{\tau}^{z} = \mathbf{C}_{\tau}^{z}$, $\tau \in \{0, \ldots, \bar{\tau}-1\}$, the same estimates as for the DKF are obtained. When the hypotheses do not match the true measurement capacities, the sum of vectors $\underline{\widetilde{\mathbf{x}}}_{k}^{s}$ yields a biased estimate $\underline{\widetilde{\mathbf{x}}}_{k}$ and must be corrected according to $(\tilde{\mathbf{\Delta}}_{k})^{-1} \underline{\widetilde{\mathbf{x}}}_{k}$ with $\tilde{\mathbf{\Delta}}_{k} = \sum_{s \in \mathcal{S}} \tilde{\mathbf{\Delta}}_{k}^{s}$. To this end, the bias processing from Section 4.1.2, i.e., (4.20) to (4.23), is utilized at each sensor, resulting in a recursive calculation of $\mathbf{\Delta}_{\tilde{k}}^{s}$ up to time step $k - \bar{\tau}$, and the calculation of a **temporary debiasing matrix** $\tilde{\mathbf{\Delta}}_{k}^{s}$ for the correction of $\underline{\widetilde{\mathbf{x}}}_{k}^{s}$.

The computational effort depends on the maximum communication delay, as for delay dependent hypotheses temporary covariances and filter gains have to be calculated. Additionally to the temporary vector $\underline{\widetilde{\mathbf{x}}}_k^s$, the HKF necessitates calculating the temporary debiasing matrix $\mathbf{\Delta}_k^s$ that emanates from the recursively obtained debiasing matrix $\mathbf{\Delta}_{k-\bar{\tau}}^s$. However, only the vector $\underline{\widetilde{\mathbf{x}}}_k^s$ and the debiasing matrix $\tilde{\mathbf{\Delta}}_k^s$ must be transmitted to the fusion center. Therefore, the communication effort is the same as for the basic HKF.

Finally, it is worth mentioning that the HKF provides unbiased estimates even if the same hypothesis is utilized for all delays at the costs of a higher MSE. Considering the **trade-off between precision and computational effort**, it is reasonable to limit the number of considered hypotheses.

For example in sensor networks where the major part of sensors transmits estimates within $\tilde{\tau}$ time steps to the fusion center and only a few sensors cause the maximum transmission delay to be $\bar{\tau} \gg \tilde{\tau}$, it makes sense to provide hypotheses only for delays $\tau \in \{0, \ldots, \tilde{\tau}-1\}$ and to use $\hat{\mathbf{C}}_{\tilde{\tau}}^z$ for all measurement capacities with $\tau \geq \tilde{\tau}$ in (4.37). Then, filter gains for $\tau \geq \tilde{\tau}$ do not change with preceding k. Thus, $\underline{\hat{\mathbf{x}}}_{\tilde{k}}^s$ and $\mathbf{\Delta}_{\tilde{k}}^s$ can be recursively calculated up to time step $k - \tilde{\tau}$. Although filter gains for delays $\tau > \tilde{\tau}$ are suboptimal in this case, the impact is small as only a few measurements are affected by the suboptimal processing, and on the other hand, the computational load is significantly lower as temporary variables have to be computed for $\tilde{\tau}$ time steps only.

Thanks to the ability to limit the number of hypotheses, the HKF is a flexible and efficient estimator for distributed systems subject to deterministic or stochastic packet delays and losses. Owing to the pre-computation of (recursive) estimates at the sensors, packet losses are compensated. Additionally, the calculation of filter gains according to sensor network capabilities optimizes the RMSE at the fusion center.

When system and measurement models are time-invariant, the necessary hypotheses in the presence of a stochastic communication can be obtained by means of

$$\hat{\mathbf{C}}^z_{k-\tilde{k}} \stackrel{!}{=} \mathrm{E}\Big\{ \sum_{s \in \mathcal{S}} \underline{\mu}^s_{k-\tilde{k}} (\mathbf{H}^s_{\tilde{k}})^\top (\mathbf{R}^s_{\tilde{k}})^{-1} \mathbf{H}^s_{\tilde{k}} \Big\} ,$$

where $\underline{\mu}^s_\tau$ denotes the random variable that indicates a transmission success between sensor s and the fusion center in the time interval τ. For time-variant measurement models, the derivation of suitable hypotheses is more elaborate as discussed next.

4.2.3 Hypothesis Optimization in Feedback Systems

Finding optimal hypotheses in general linear systems is an optimization problem by itself. Given the stochastic attributes of the communication network and a model about the evolution of the measurement capacity, the measurement capacities can be estimated and the hypotheses that minimize the expected covariance of the estimate at the fusion center can be calculated. Unfortunately, the evolution of the measurement capacity is usually state-dependent, nonlinear, and unknown so that such a processing is not only complex but also quite often impossible to realize.

The approach pursued in this section is to estimate measurement capacities from past time steps based on the fused debiasing matrix at the fusion center. The estimates of the measurement capacities are then used as new hypotheses and are communicated back to the sensors in a **feedback setup**.

In order to facilitate the estimation of the measurement capacity subject to complex evolution models, the derivations are based on the assumption that the measurement capacities as well as the state model are (approximately) time-invariant.

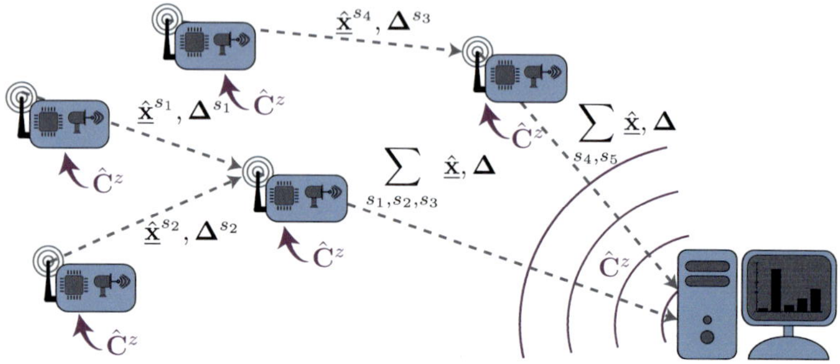

Figure 4.9: The HKF with the proposed feedback extension. The fusion center collects debiasing matrices from the sensors, obtains the average true measurement capacity from past time steps, calculates a new hypothesis $\hat{\mathbf{C}}^z$, and broadcasts it. The sensors receive the new hypothesis and use it for the calculation of local covariances $\hat{\mathbf{C}}$.

Moreover, for the sake of simplicity, the considerations are confined to sensor networks without time delays. In the following, key formulas are presented. An extensive discussion and evaluation of the approach is provided in [185].

It is well known that for time-invariant systems, the KF covariance converges to a **steady state**. Consider the global covariance of the HKF from (4.25) to (4.27) for time-invariant state matrices that corresponds to the covariance of a KF with measurement model $\hat{\mathbf{C}}^z$. When there exist matrices $\mathbf{H}$, $\mathbf{R}$ with

$$\hat{\mathbf{C}}^z = (\mathbf{H})^\top (\mathbf{R})^{-1} \mathbf{H}$$

and $(\mathbf{H}, \mathbf{A})$ is observable according to Definition 2.2, the KF with measurement capacity $\hat{\mathbf{C}}^z$ is stable [83]. Therefore, $\hat{\mathbf{C}}_k$ converges to a steady state $\hat{\mathbf{C}}_{ss}$. From the steady state covariance, constant filter gains ensue.

Now, the calculation of the debiasing matrix is examined for time-invariant covariances and gains. Writing out the calculations of $\mathbf{\Delta}_k$

by means of (4.21) to (4.23) amounts to

$$\begin{aligned}\boldsymbol{\Delta}_k &= \sum_{s\in\mathcal{S}} \boldsymbol{\Delta}^s_{k|k} \\ &= \sum_{s\in\mathcal{S}} (\mathbf{L}\mathbf{A}\boldsymbol{\Delta}^s_{k-1}(\mathbf{A})^{-1} + \mathbf{K}^s\mathbf{H}^s) \\ &= \mathbf{L}\mathbf{A}\Big(\sum_{s\in\mathcal{S}} \boldsymbol{\Delta}^s_{k-1}\Big)(\mathbf{A})^{-1} + \sum_{s\in\mathcal{S}} \mathbf{K}^s\mathbf{H}^s \ .\end{aligned}$$

A repeated application of the decomposition leads to a "rolled out" description of the debiasing matrix

$$\boldsymbol{\Delta}_k = \sum_{\tilde{k}=1}^{k} (\mathbf{L}\mathbf{A})^{k-\tilde{k}} \Big(\sum_{s\in\mathcal{S}} \mathbf{K}^s\mathbf{H}^s\Big)\mathbf{A}^{\tilde{k}-k} + (\mathbf{L}\mathbf{A})^k \sum_{s\in\mathcal{S}} \mathbf{T}^s\mathbf{A}^{-k} \ .$$

When $\boldsymbol{\Delta}_k$ converges for $k \to \infty$, the last term can be neglected so that the formula simplifies by means of $\sum_{s\in\mathcal{S}} \mathbf{K}^s\mathbf{H}^s = \hat{\mathbf{C}}_{ss}\mathbf{C}^z$ to

$$\boldsymbol{\Delta}_k = \sum_{\tilde{k}=1}^{k} (\mathbf{L}\mathbf{A})^{k-\tilde{k}} \hat{\mathbf{C}}_{ss}\mathbf{C}^z\mathbf{A}^{\tilde{k}-k} \ . \tag{4.38}$$

Hence, (4.38) describes the functional dependency between known matrices and unknown measurement capacity $\mathbf{C}^z$. Utilizing Kronecker product and vector operator [130], (4.38) is transformed to

$$\mathrm{vec}\{\boldsymbol{\Delta}_k\} = \Big(\sum_{\tilde{k}=1}^{k} (\mathbf{A}^{k-\tilde{k}})^{-\top} \otimes (\mathbf{L}\mathbf{A})^{k-\tilde{k}}\Big)\, \mathrm{vec}\{\hat{\mathbf{C}}_{ss}\mathbf{C}^z\} \ .$$

It follows with $(\mathbf{A}^{k-\tilde{k}})^{-\top} \otimes (\mathbf{L}\mathbf{A})^{k-\tilde{k}} = \big((\mathbf{A})^{-\top} \otimes \mathbf{L}\mathbf{A}\big)^{k-\tilde{k}}$ that

$$\begin{aligned}\mathrm{vec}\{\boldsymbol{\Delta}_k\} &= \sum_{\tilde{k}=1}^{k} \big((\mathbf{A})^{-\top} \otimes \mathbf{L}\mathbf{A}\big)^{k-\tilde{k}}\, \mathrm{vec}\{\hat{\mathbf{C}}_{ss}\mathbf{C}^z\} \\ &= \sum_{\tilde{k}=0}^{k-1} \big((\mathbf{A})^{-\top} \otimes \mathbf{L}\mathbf{A}\big)^{\tilde{k}}\, \mathrm{vec}\{\hat{\mathbf{C}}_{ss}\mathbf{C}^z\} \ .\end{aligned}$$

For $k \to \infty$, the sum term corresponds to the Neumann series [130]. Hence, when the eigenvalues $\operatorname{eig}\{(\mathbf{A})^{-\top} \otimes \mathbf{LA}\} = \operatorname{eig}\{(\mathbf{A})^{-\top}\} \otimes \operatorname{eig}\{\mathbf{LA}\}$ are in the unit circle, the sum term corresponds to

$$\operatorname{vec}\{\mathbf{\Delta}_k\} = (\mathbf{I} - (\mathbf{A})^{-\top} \otimes \mathbf{LA})^{-1} \operatorname{vec}\{\hat{\mathbf{C}}_{ss}\mathbf{C}^z\} \ .$$

This, finally, leads to

$$\begin{aligned}\operatorname{vec}\{\hat{\mathbf{C}}_{ss}\mathbf{C}^z\} &= (\mathbf{I} - (\mathbf{A})^{-\top} \otimes \mathbf{LA}) \operatorname{vec}\{\mathbf{\Delta}_k\} \\ &= \operatorname{vec}\{\mathbf{\Delta}_k - \mathbf{LA}\mathbf{\Delta}_k(\mathbf{A})^{-1}\} \ .\end{aligned}$$

Theorem 4.5 *Let* $\mathbf{L}$ *denote the gain* (4.28) *obtained with the steady state covariance* $\hat{\mathbf{C}}_{ss}$ *of a KF with measurement capacity* $\hat{\mathbf{C}}^z$*, and let* $\mathbf{\Delta}_{ss}$ *denote the corresponding debiasing matrix. For observable time-invariant linear systems according to Definition 2.1 and*

$$\max |\operatorname{eig}\{(\mathbf{A})^{-\top}\} \otimes \operatorname{eig}\{\mathbf{LA}\}| < 1 \ ,$$

the measurement capacity is given by

$$\mathbf{C}^z = (\hat{\mathbf{C}}_{ss})^{-1}(\mathbf{\Delta}_{ss} - \mathbf{LA}\mathbf{\Delta}_{ss}(\mathbf{A})^{-1}) \ . \tag{4.39}$$

The covariance $\hat{\mathbf{C}}_{ss}$ is the solution of the (discrete-time) algebraic Riccati equation [129] that is obtained with state model matrices and hypothesis $\hat{\mathbf{C}}^z$ along with the gain $\mathbf{L}$ from (4.28).

Therefore, the feedback hypothesis can be calculated at the fusion center based only on the fused debiasing matrix that is transmitted from the sensors anyway. A schematic overview of the communication is given in Figure 4.9.

By employing the proposed feedback technique, the HKF provides estimates that are asymptotically equal to the LMMSE estimates for time-invariant systems. For an evaluation of the scheme in time-variant systems, see [185]. Indeed, ensuring that the same hypothesis is used at all sensors is not always feasible. For such scenarios, the more general concept presented in the next section provides a solution.

4.3 Generalized Hypothesizing Filtering

DKF and HKF optimize filter gains according to a global measurement capacity. By this means, they provide higher precision at the fusion center than methods that are based on local KFs. So far, it was assumed that all sensors employ the same hypothesis about the measurement capacity. This turned out to be a meaningful requirement to derive simple fusion rules and to obtain a feedback technique for the optimization of the hypothesis.

However, a hypothesis that is network-wide known is restrictive, as it must be asserted that a change of the hypothesis, e.g., as proposed in Section 4.2.3, is acknowledged by all sensors in the network. Consider a temporary communication failure due to which a new hypothesis is not received by one of the sensors. If no control mechanism is established, the fusion center estimate becomes suboptimal and the feedback technique does not work anymore.

Therefore, either, the hypothesis should not be changed then, or all packets received by the fusion center must be checked for compliance with the transmitted hypotheses as it has been proposed in [185]. For both solutions, an overhead in planning and implementation is necessary while the additive fusion rule is suboptimal when the assumptions of Corollary 4.4 are not satisfied.

In this section, the filter gain optimization according to a global measurement model is exploited more generally by letting sensors employ **individual hypotheses**. Indeed, the ideas derived in this section are not to be understood as a plug-and-play estimator but as a framework for the efficient optimization of filter gains in sensor networks.

- **Covariance bounds** for the fused estimate are derived based on sensor variables that are calculated without global model knowledge.

- Based on covariance bounds, an efficient **fusion rule** that is applicable to the generalized HKF concept is proposed.
- Inflated local measurement models are used as substitutes for the global measurement model in the derivation of the **inflation Kalman filter**.

The results are generalizations of the HKF that, in particular, apply to the theory from the last sections. Note also that the debiasing technique from Section 4.1.2 does not rely on identical hypotheses. Therefore, it is directly applicable to distributed estimation with differing hypotheses.

4.3.1 Covariance Bounds for Hypothesizing Estimation

The centralized KF provides closed-form equations for the calculation of covariances, which were presented in the context of the DKF (4.4) and (4.6). For centralized unbiased linear estimators, the covariance is obtained according to (2.22) and (2.24). However, as discussed in Chapter 3, the common process noise causes dependencies between estimates in distributed systems that must be considered in the derivation of the covariance of the fused estimate.

For the HKF, different techniques have been proposed to obtain covariances for the fused estimate. In [184], the recursive calculation of auxiliary matrices was presented that enables the derivation of covariance bounds at the fusion center. The main idea was to differentiate between (dependent) process noise and (independent) measurement noise terms. As the bound becomes conservative when the measurement noise covariances differ, an approximation based on a steady state analysis was proposed in [185] that aims to reflect the true covariance instead of providing a bound. A different path was taken in [55], where the authors proposed a double-debiasing technique that provides a consistent estimate of the true covariance.

In the sequel of this section, the bounding approach from [184] is improved. The technique is generalized to cope with different hypotheses and a tighter bound is found by applying the bounding technique from Section 3.2.

For this purpose, let each sensor maintain two matrices that account for independent and dependent noise terms respectively. Let these auxiliary covariances be denoted by $\mathbf{B}^{id}$ and $\mathbf{B}^{d}$. Assuming independent initial estimates, separating measurement and process noise, and applying the linear transformations of the HKF to the auxiliary variables amounts to the following Lemmata.

Lemma 4.6 *Let* $\hat{\underline{\mathbf{x}}}$ *denote an estimate obtained with Algorithm 2.1 and let* $\mathbf{\Delta}$ *denote the corresponding debiasing matrix. The covariance of the estimate* $\tilde{\underline{\mathbf{x}}} = (\mathbf{\Delta})^{-1}\hat{\underline{\mathbf{x}}}$ *is given by*

$$\mathbf{C} = \mathrm{E}\{(\tilde{\underline{\mathbf{x}}} - \underline{\mathbf{x}})^2\} = (\mathbf{\Delta})^{-1}(\mathbf{B}^{id} + \mathbf{B}^{d})(\mathbf{\Delta})^{-\top} , \tag{4.40}$$

where $\mathbf{B}^{id}$ *and* $\mathbf{B}^{d}$ *are recursively obtained* [4] *according to*

$$\textit{Initialization: } \mathbf{B}_0^{id} = \mathbf{\Delta}_0 \mathbf{C}_0 (\mathbf{\Delta}_0)^\top , \tag{4.41}$$

$$\textit{Prediction: } \mathbf{B}_{k+1}^{id} = \mathbf{A}_k \mathbf{B}_k^{id} (\mathbf{A}_k)^\top , \tag{4.42}$$

$$\textit{Filtering: } \mathbf{B}_{k|k}^{id} = \mathbf{L}_k \mathbf{B}_k^{id} (\mathbf{L}_k)^\top + \mathbf{K}_k \mathbf{R}_k (\mathbf{K}_k)^\top , \tag{4.43}$$

and

$$\textit{Initialization: } \mathbf{B}_0^{d} = \mathbf{0} , \tag{4.44}$$

$$\textit{Prediction: } \mathbf{B}_{k+1}^{d} = \mathbf{A}_k \mathbf{B}_k^{d} (\mathbf{A}_k)^\top + \mathbf{\Delta}_{k+1} \mathbf{Q}_k (\mathbf{\Delta}_{k+1})^\top , \tag{4.45}$$

$$\textit{Filtering: } \mathbf{B}_{k|k}^{d} = \mathbf{L}_k \mathbf{B}_k^{d} (\mathbf{L}_k)^\top . \tag{4.46}$$

[4] The formulas are given for the initialization with independent estimates. When the sensors are initialized with the same estimate, $\mathbf{\Delta}_0 \mathbf{C}_0 (\mathbf{\Delta}_0)^\top$ must be added to $\mathbf{B}_0^{d}$ and not to $\mathbf{B}_0^{id}$.

PROOF. According to Theorem 4.2, $\underline{\widetilde{\mathbf{x}}} = (\boldsymbol{\Delta})^{-1}\underline{\hat{\mathbf{x}}}$ is unbiased. Thus, the true covariance is given by the equations for unbiased linear filters (2.22) and (2.24). The claim is obviously fulfilled at initialization. In the prediction step, it holds

$$\begin{aligned}\mathbf{C}_{k+1} =& \mathbf{A}_k\mathbf{C}_k(\mathbf{A}_k)^\top + \mathbf{Q}_k \\ =& \boldsymbol{\Delta}_{k+1}^{-1}(\mathbf{A}_k\boldsymbol{\Delta}_k\mathbf{C}_k\boldsymbol{\Delta}_k^\top(\mathbf{A}_k)^\top + \boldsymbol{\Delta}_{k+1}\mathbf{Q}_k\boldsymbol{\Delta}_{k+1}^\top)\boldsymbol{\Delta}_{k+1}^{-\top} ,\end{aligned}$$

which proves correctness with

$$\mathbf{A}_k\boldsymbol{\Delta}_k\mathbf{C}_k(\boldsymbol{\Delta}_k)^\top(\mathbf{A}_k)^\top = \mathbf{A}_k\mathbf{B}_k^{id}(\mathbf{A}_k)^\top + \mathbf{A}_k\mathbf{B}_k^{d}(\mathbf{A}_k)^\top .$$

In the filter step, the transformation matrices are given by

$$\tilde{\mathbf{L}}_k =(\boldsymbol{\Delta}_{k|k})^{-1}\mathbf{L}_k\boldsymbol{\Delta}_k \text{ and } \tilde{\mathbf{K}}_k = (\boldsymbol{\Delta}_{k|k})^{-1}\mathbf{K}_k ,$$

so that the covariance is given by

$$\begin{aligned}\mathbf{C}_{k|k} =& \boldsymbol{\Delta}_{k|k}^{-1}(\mathbf{L}_k\boldsymbol{\Delta}_k\mathbf{C}_k\boldsymbol{\Delta}_k^\top(\mathbf{L}_k)^\top + \mathbf{K}_k\mathbf{R}_k(\mathbf{K}_k)^\top)\boldsymbol{\Delta}_{k|k}^{-\top} \\ =& \boldsymbol{\Delta}_{k|k}^{-1}(\mathbf{L}_k\mathbf{B}_k^{id}(\mathbf{L}_k)^\top + \mathbf{L}_k\mathbf{B}_k^{d}(\mathbf{L}_k)^\top + \mathbf{K}_k\mathbf{R}_k(\mathbf{K}_k)^\top)\boldsymbol{\Delta}_{k|k}^{-\top} ,\end{aligned}$$

which concludes the proof. □

Lemma 4.7 *Let the assumptions be the same as in Lemma 4.6. The cross-covariance matrix between estimates* $\underline{\widetilde{\mathbf{x}}}_k^s = (\boldsymbol{\Delta}_k^s)^{-1}\underline{\hat{\mathbf{x}}}_k^s$ *and* $\underline{\widetilde{\mathbf{x}}}_k^{\tilde{s}} = (\boldsymbol{\Delta}_k^{\tilde{s}})^{-1}\underline{\hat{\mathbf{x}}}_k^{\tilde{s}}$ *is given by*

$$\mathbf{C}_k^{s\tilde{s}} = \mathrm{E}\{(\underline{\widetilde{\mathbf{x}}}_k^s - \underline{\mathbf{x}}_k)(\underline{\widetilde{\mathbf{x}}}_k^{\tilde{s}} - \underline{\mathbf{x}}_k)^\top\} = (\boldsymbol{\Delta}_k^s)^{-1}\mathbf{B}_k^{s\tilde{s}_d}(\boldsymbol{\Delta}_k^{\tilde{s}})^{-\top} ,$$

where $\mathbf{B}_k^{s\tilde{s}_d}$ *is recursively obtained according to*

$$\begin{aligned}\textit{Initialization: } \mathbf{B}_0^{s\tilde{s}_d} =& \mathbf{0}\ ^5 && (4.47)\\ \textit{Prediction: } \mathbf{B}_{k+1}^{s\tilde{s}_d} =& \mathbf{A}_k\mathbf{B}_k^{s\tilde{s}_d}(\mathbf{A}_k)^\top + \boldsymbol{\Delta}_{k+1}^s\mathbf{Q}_k(\boldsymbol{\Delta}_{k+1}^{\tilde{s}})^\top , && (4.48)\\ \textit{Filtering: } \mathbf{B}_{k|k}^{s\tilde{s}_d} =& \mathbf{L}_k^s\mathbf{B}_k^{s\tilde{s}_d}(\mathbf{L}_k^{\tilde{s}})^\top . && (4.49)\end{aligned}$$

PROOF. As the measurement noise of remote nodes is independent, it holds

$$\mathrm{E}\{(\hat{\underline{\mathbf{x}}}_k^s - \underline{\mathbf{x}}_k)(\underline{\mathbf{v}}_k^{\tilde{s}})^\top\} = \mathbf{0}$$

and

$$\mathrm{E}\{\underline{\mathbf{v}}_k^s(\hat{\underline{\mathbf{x}}}_k^{\tilde{s}} - \underline{\mathbf{x}}_k)^\top\} = \mathbf{0} \ .$$

Thus, the cross-covariance matrix of the filtered estimates is given by

$$\mathbf{C}_{k|k}^{s\tilde{s}} = (\mathbf{\Delta}_{k|k}^s)^{-1}\mathbf{L}_k^s\mathbf{\Delta}_k^s\mathbf{C}_k^{s\tilde{s}}(\mathbf{\Delta}_k^{\tilde{s}})^\top(\mathbf{L}_k^{\tilde{s}})^\top(\mathbf{\Delta}_{k|k}^{\tilde{s}})^{-\top} \ .$$

The remaining part of the proof follows along the lines of the proof of Lemma 4.6. □

For known cross-covariance matrices, the covariance of the fused estimate is obtained with (2.27). However, as shown in Lemma 4.7, cross-covariance matrices depend on transformation and debiasing gains of remote nodes. Therefore, global model knowledge is required.

In [184], it is demonstrated that a bound of the fused estimate is achieved by equally inflating local process noise parts. Indeed, tighter bounds are obtained with the inflation techniques from Section 3.2.

Theorem 4.8 *Let* $\hat{\underline{\mathbf{x}}}^s$ *denote estimates obtained by means of Algorithm 2.1 and let* $\mathbf{\Delta}^s$ *denote the corresponding debiasing matrices. The covariance of the fused estimate*

$$\tilde{\underline{\mathbf{x}}} = (\mathbf{\Delta})^{-1}\hat{\underline{\mathbf{x}}} \ , \ \textit{where} \ \hat{\underline{\mathbf{x}}} = \sum_{s\in\mathcal{S}} \mathbf{F}^s\hat{\underline{\mathbf{x}}}^s \qquad (4.50)$$

with debiasing matrix $\mathbf{\Delta}$ *from* (4.23) *is bounded by*

$$\mathbf{C} \leq (\mathbf{\Delta})^{-1}\sum_{s\in\mathcal{S}} \mathbf{F}^s\Big(\mathbf{B}^{s_{id}} + \frac{1}{\omega^s}\mathbf{B}^{s_d}\Big)(\mathbf{F}^s)^\top(\mathbf{\Delta})^{-\top} \ , \qquad (4.51)$$

with $\sum_{s\in\mathcal{S}}\omega^s = 1$, $\omega^s > 0$.

PROOF. Considering the unbiased local estimates $\underline{\widetilde{\mathbf{x}}}^s = (\mathbf{\Delta}^s)^{-1}\underline{\hat{\mathbf{x}}}^s$, the fusion operation is given by

$$\widetilde{\mathbf{F}}^s = (\mathbf{\Delta})^{-1}\mathbf{F}^s\mathbf{\Delta}^s \; .$$

Thus, it holds,

$$\mathbf{C} = (\mathbf{\Delta})^{-1}\Big(\sum_{s,\tilde{s}\in\mathcal{S}} \mathbf{F}^s\mathbf{\Delta}^s\mathbf{C}^{s\tilde{s}}(\mathbf{\Delta}^{\tilde{s}})^\top(\mathbf{F}^{\tilde{s}})^\top\Big)(\mathbf{\Delta})^{-\top} \; ,$$

where $\mathbf{C}^{s\tilde{s}}$ denote the (cross-)covariance matrices between unbiased estimates $\underline{\widetilde{\mathbf{x}}}^s$ and $\underline{\widetilde{\mathbf{x}}}^{\tilde{s}}$. From Lemmata 4.6 and 4.7, it follows that

$$\mathbf{\Delta}^s\mathbf{C}^{s\tilde{s}}(\mathbf{\Delta}^{\tilde{s}})^\top = \begin{cases} \mathbf{B}^{s_{id}} + \mathbf{B}^{s_d} & \text{for } s = \tilde{s} \\ \mathbf{B}^{s\tilde{s}_d} & \text{else} \end{cases} \; .$$

As sum of linearly transformed positive semi-definite matrices, the joint covariance matrix of the dependent parts $\mathbf{B}^{s_d}$ and $\mathbf{B}_k^{s\tilde{s}_d}$ is positive semi-definite. From Lemma 3.9 and Lemma 3.12, it follows that

$$\sum_{s,\tilde{s}\in\mathcal{S}} \mathbf{F}^s\mathbf{C}^{s\tilde{s}_d}(\mathbf{F}^{\tilde{s}})^\top \leq \sum_{s\in\mathcal{S}} \mathbf{F}^s\frac{1}{\omega^s}\mathbf{C}^{s_d}(\mathbf{F}^{\tilde{s}})^\top \; ,$$

which concludes the proof. □

The parameters $\{\omega^s\}_{s\in\mathcal{S}}$ are chosen to optimize a criterion such as the trace or determinant of the bound. A detailed discussion on properties of this optimization is given in Section 3.2.2. Furthermore, it is worth mentioning that the bound presented in [184] is a special case of Theorem 4.8, namely $\omega^s = \frac{1}{S}$ and $\mathbf{F}_k^s = \mathbf{I}$.

The **algorithm to obtain a covariance bound** at the fusion center by means of Theorem 4.8 consists of sensor and fusion center processing. At the sensors, the matrices $\mathbf{B}^{s_d}$ and $\mathbf{B}^{s_{id}}$ are recursively calculated according to Lemma 4.6. Then, both matrices are transmitted along with the estimate and the debiasing matrix to the fusion center. Note that $\mathbf{B}^{s_d}$ and $\mathbf{B}^{s_{id}}$ are symmetric so that only triangular matrices must be communicated. At the fusion center, the estimates are combined with an arbitrary fusion rule and a bound is calculated with Theorem 4.8. The necessary operations are schematically illustrated in Figure 4.10.

4.3.2 Fusion Methods for Biased Estimates

The fusion rule for the HKF is given as the sum operation of biased estimates in (4.30) and (4.31). Considering the unbiased counterparts $\underline{\widetilde{\mathbf{x}}}^s = (\mathbf{\Delta}^s)^{-1}\underline{\hat{\mathbf{x}}}^s$, the fusion operation (4.30) is a weighted combination

$$\underline{\widetilde{\mathbf{x}}}_k = \sum_{s \in \mathcal{S}} \mathbf{F}^s \underline{\hat{\mathbf{x}}}^s \text{ with } \mathbf{F}^s = (\mathbf{\Delta})^{-1}\mathbf{\Delta}^s \,.$$

The matrix weights $\mathbf{F}^s$ are determined by biases of local estimates. More precisely, when the same hypothesis is used at the sensors, identical gains $\mathbf{L}_k$ and the same global covariance are obtained, which, in turn, lead to debiasing matrices that depend on the relation between local measurement models and global measurement capacity. When the hypothesis about the global measurement capacity is correct, it is a consequence of Corollary 4.4 that the fusion gains correspond to the LMMSE combination from Theorem 2.5. Otherwise, the technique can fail as demonstrated in the next example.

Example 4.6: Suboptimal Fusion Rules for the Generalized HKF

Consider a sensor network consisting of two identical sensors. Let node s_1 apply a hypothesis that underestimates the true measurement capacity by 50 percent and node s_2 a hypothesis that overestimates it by 100 percent. Then, a weighting based on debiasing matrices would allot the unbiased estimate at node s_1 four times more weight than the other one although both estimates carry the same amount of information. ■

Even though Example 4.6 is not realistic as the local filter processing is to be optimized according to meaningful hypotheses, the underlying problem should be resolved. As a matter of fact, the HKF and its generalization are linear estimators. Therefore, the LMMSE combination rule from Theorem 2.5 provides the best fusion results. However, as discussed before, the LMMSE combination necessitates the calculation of cross-covariance matrices. Hence it is only applicable when elaborate reconstruction techniques are used.

In this regard, the additive fusion rule of the basic HKF is a compromise that accepts suboptimal precision in order to keep complexity and computational effort limited. Indeed, for the generalization with differing hypotheses, the fusion rule is no longer justified as illustrated in Example 4.6. Therefore, the bounding techniques from Chapter 3 are adapted to the HKF processing as follows.

Theorem 4.9 *Let* $\mathbf{B}^{s_{id}}$ *and* $\mathbf{B}^{s_d}$, $s \in \mathcal{S}$ *denote the matrices from Lemma 4.6. Then, the fusion gains that minimize the bound from Theorem 4.8 are given by*

$$\mathbf{F}^s = \mathbf{P}(\boldsymbol{\Delta}^s)^\top \left(\mathbf{B}^{s_{id}} + \frac{1}{\omega^s}\mathbf{B}^{s_d}\right)^{-1} \tag{4.52}$$

with

$$\mathbf{P} = \left(\sum_{s\in\mathcal{S}} (\boldsymbol{\Delta}^s)^\top \left(\mathbf{B}^{s_{id}} + \frac{1}{\omega^s}\mathbf{B}^{s_d}\right)^{-1} \boldsymbol{\Delta}^s\right)^{-1} , \tag{4.53}$$

where $\{\omega^s\}_{s\in\mathcal{S}}$ *are the solutions to the constrained convex optimization problem* $\arg\min_{\omega^s, s\in\mathcal{S}} \mathbf{P}$ *with* $\sum_{s\in\mathcal{S}} \omega^s = 1$ *and* $\omega^s > 0$.

PROOF. The debiasing matrix $\boldsymbol{\Delta}$ from Theorem 4.8 accounts for a possible bias of the fused estimate and, without loss of generality, can be integrated into the optimization of fusion gains $\mathbf{F}^s$. An unbiased estimate, in turn, is yielded by correcting $\underline{\hat{\mathbf{x}}}^s$ with $(\boldsymbol{\Delta}^s)^{-1}$ and fusing the estimates with $\sum_{s\in\mathcal{S}} \tilde{\mathbf{F}}^s = \mathbf{I}$, i.e., $\mathbf{F}^s = \tilde{\mathbf{F}}(\mathbf{\Delta}^s)^{-1}$. Therefore, the considered problem can be posed as minimizing

$$\sum_{s\in\mathcal{S}} \tilde{\mathbf{F}}^s(\mathbf{\Delta}^s)^{-1}\left(\mathbf{B}^{s_{id}} + \frac{1}{\omega^s}\mathbf{B}^{s_d}\right)(\mathbf{\Delta}^s)^{-\top}(\tilde{\mathbf{F}}^s)^\top$$

subject to $\sum_{s\in\mathcal{S}} \tilde{\mathbf{F}}^s = \mathbf{I}$. This problem, however, is a special case of Theorem 3.14 with block diagonal joint covariance matrix. Expanding the minimization leads to (4.53) so that the optimal weights are given by

$$\tilde{\mathbf{F}}^s = \mathbf{P}(\mathbf{\Delta}^s)^\top\left(\mathbf{B}^{s_{id}} + \frac{1}{\omega^s}\mathbf{B}^{s_d}\right)^{-1}\mathbf{\Delta}^s\,,$$

which eventually results in (4.52) with $\mathbf{F}^s = \tilde{\mathbf{F}}(\mathbf{\Delta}^s)^{-1}$. The convexity of the optimization follows with Theorem 3.15. □

Extensions and properties of Theorem 4.9 follow along the lines of Section 3.2. For example, it is possible to reduce the computational effort by replacing the optimization of the weights $\{\omega^s\}_{s\in\mathcal{S}}$ by heuristics [48, 118]. This makes sense, as the computational effort at the fusion center is higher than for the basic HKF where only summations and one inversion in state dimension are necessary.

In the framework of the generalized HKF, the filter processing at the nodes is optimized according to **local hypotheses**. To this end, sensors optimize local filter gains by means of the HKF formulas (4.25) to (4.27). The local hypothesis about the global measurement capacity can be construed as a "best guess" at the sensor. The fusion center uses the estimates and matrices from the sensors to obtain optimized fusion gains with Theorem 4.9. The processing is illustrated in Figure 4.10. In particular, it is viable to process biased estimates at the sensors and to derive covariance bounds.

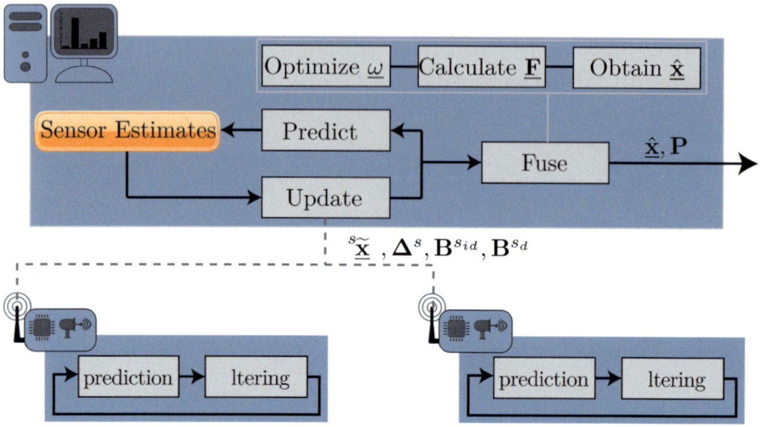

Figure 4.10: The processing scheme of the generalized HKF. The sensors process $\hat{\underline{\mathbf{x}}}^{s}, \mathbf{\Delta}^{s}, \mathbf{B}^{sid}, \mathbf{B}^{sd}$ locally and send the variables to the fusion center. The fusion center collects all data and computes the fused estimate by means of Theorem 4.9.

4.3.3 Inflation Kalman Filter

The generalized HKF in combination with the debiasing technique from Section 4.1.2 provides a high degree of freedom in the filter design. However, it is not always feasible to employ meaningful best guesses about the measurement capacity at the sensors or the complexity of the algorithm can be too high for a considered problem. For such cases, a simple algorithm that inflates local measurement models for the optimization of filter gains can be used.

As motivation consider a **homogeneous sensor network**, i.e., a network consisting of identical sensors that feature the same measurement model (2.9). Then, the measurement capacity simplifies to

$$\mathbf{C}^{z} = \sum_{s \in \mathcal{S}} (\mathbf{H})^{\top} (\mathbf{R})^{-1} \mathbf{H} = \gamma (\mathbf{H})^{\top} (\mathbf{R})^{-1} \mathbf{H} \ ,$$

where $\gamma = |\mathcal{S}|$ denotes a scalar factor. Note that the equality still holds when the measurement noise covariances at the sensors are scalar multitudes of a common matrix $\mathbf{R}$. The idea is to hypothesize the scaling factor and to inflate local sensor models $(\mathbf{H})^\top(\mathbf{R})^{-1}\mathbf{H}$ with γ to approximate the measurement capacity. By means of the measurement capacity, **global covariances** are recursively obtained according to

$$\text{Initialization: } \hat{\mathbf{C}}_0 = \frac{1}{\gamma}\mathbf{C}_0 \,, \tag{4.54}$$

$$\text{Prediction: } \hat{\mathbf{C}}_{k+1} = \mathbf{A}_k\hat{\mathbf{C}}_k(\mathbf{A}_k)^\top + \mathbf{Q}_k \,, \tag{4.55}$$

$$\text{Filtering: } \hat{\mathbf{C}}_{k|k} = \left((\hat{\mathbf{C}}_k)^{-1} + \gamma(\mathbf{H}_k)^\top(\mathbf{R}_k)^{-1}\mathbf{H}_k\right)^{-1} . \tag{4.56}$$

So far, the algorithm corresponds to the generalized HKF with hypotheses

$$\hat{\mathbf{C}}^z = \gamma(\mathbf{H})^\top(\mathbf{R})^{-1}\mathbf{H} \,.$$

However, in order to reduce computational and communication effort, filter gains are enforced to be unbiased. For this purpose, the filter gain $\mathbf{K}$ is inflated with γ, yielding the formulas

$$\mathbf{L}_k = \hat{\mathbf{C}}_{k|k}(\hat{\mathbf{C}}_k)^{-1} \,, \tag{4.57}$$

$$\mathbf{K}_k = \gamma\hat{\mathbf{C}}_{k|k}(\mathbf{H}_k)^\top(\mathbf{R}_k)^{-1} \,. \tag{4.58}$$

Then, it holds

$$\begin{aligned}\mathbf{I} - \mathbf{K}_k\mathbf{H}_k &= \hat{\mathbf{C}}_{k|k}(\hat{\mathbf{C}}_{k|k})^{-1} - \gamma\hat{\mathbf{C}}_{k|k}(\mathbf{H}_k)^\top(\mathbf{R}_k)^{-1}\mathbf{H}_k \\ &= \hat{\mathbf{C}}_{k|k}(\hat{\mathbf{C}}_k)^{-1} \\ &= \mathbf{L}_k^s \,.\end{aligned}$$

Therefore, the filter gains are unbiased according to (2.25). Moreover, the central KF estimate can be reconstructed in homogeneous sensor networks.

Theorem 4.10 *Let the initial estimates* $\widetilde{\underline{\mathbf{x}}}_0^s$ *be independent and subject to a common uncertainty* $\mathbf{C}_0^s$*, and let the measurement model be the same for all sensors* $s \in \mathcal{S}$*. When the sensors apply Algorithm 2.1 with* $\mathbf{T} = \mathbf{I}$ *and filter gains* (4.57) *and* (4.58) *with* $\gamma = |\mathcal{S}|$*, the fusion center estimate*

$$\hat{\underline{\mathbf{x}}}_k = \frac{1}{|\mathcal{S}|} \sum_{s \in \mathcal{S}} \hat{\underline{\mathbf{x}}}_k^s \tag{4.59}$$

corresponds to the LMMSE estimate.

Proof. Let the central KF covariance be denoted by $\mathbf{C}$. It is proven that $\hat{\mathbf{C}} = \mathbf{C}$ and that the transformation matrices of the inflation KF correspond to the ones of the centralized KF. At initialization, the covariance $\hat{\mathbf{C}}_0$ reflects the central KF covariance

$$\mathbf{C}_0 = (\sum_{s \in \mathcal{S}} (\mathbf{C}_0^s)^{-1})^{-1} = \frac{1}{|\mathcal{S}|} \mathbf{C}_0^s$$

and the LMMSE estimate is given by

$$\mathbf{C}_0 (\mathbf{C}_0^s)^{-1} = \frac{1}{|\mathcal{S}|} \sum_{s \in \mathcal{S}} \widetilde{\underline{\mathbf{x}}}_k^s \ .$$

In the prediction step, it holds $\hat{\mathbf{C}}_{k+1} = \mathbf{C}_{k+1}$ when $\hat{\mathbf{C}}_k = \mathbf{C}_k$ and the transformations of the estimate in centralized KF and inflation KF are the same. In the filter step, the central KF covariance is given by

$$\begin{aligned}
\mathbf{C}_{k|k} &= \Big((\mathbf{C}_k)^{-1} + \sum_{s \in \mathcal{S}} (\mathbf{H}_k^s)^\top (\mathbf{R}_k^s)^{-1} \mathbf{H}_k^s \Big)^{-1} \\
&= \Big((\mathbf{C}_k)^{-1} + |\mathcal{S}| (\mathbf{H}_k^s)^\top (\mathbf{R}_k^s)^{-1} \mathbf{H}_k^s \Big)^{-1} = \hat{\mathbf{C}}_{k|k} \ .
\end{aligned}$$

Therefore, the gains $\mathbf{L}$ are the same for both estimators. Furthermore, the factor γ in (4.58) is canceled out due to fusion formula $\hat{\underline{\mathbf{x}}} = \frac{1}{|\mathcal{S}|} \sum_{s \in \mathcal{S}} \hat{\underline{\mathbf{x}}}$, which concludes the proof. □

As local estimates obtained with the inflation KF are unbiased and the true covariance of the estimates is the same for all nodes in homogeneous sensor networks, the natural fusion operation is to take the average (4.59). This also applies when not all estimates are received at the fusion center. For **heterogeneous sensor networks**, the variable γ does not need to be the same for all sensors. In particular, it is reasonable to optimize the value of γ according to the neighborhood of sensors. In fact, the localized optimization amounts to formulas similar to the ones of the consensus Kalman filter [125].

Note also that the processing of the inflation KF is closely related to the **federated Kalman filter** [24, 25]. However, the federated Kalman filter achieves bounds on the fused estimate by inflating the process noise covariance instead of the measurement model. Thus, suboptimal filter gains are obtained and the LMMSE estimate cannot be reconstructed at the fusion center after arbitrary many time steps.

It is also worth pointing out that bounds and more efficient fusion rules can be obtained with the techniques from Chapter 3 and Section 4.3.1. In particular, the bound from Theorem 4.8 is tight for homogeneous sensor networks as shown in [184].

The main advantage of the inflation KF is its simplicity and the possibility to derive unbiased estimates at the sensors that are optimized according to the measurement capacity of the sensor network. However, when the sensor models differ, e.g., one half of the sensors observes the position and the other half the velocity of a target, the inflation yields suboptimal filter gains as any multitude of the local sensor models is a bad approximation of the true measurement capacity. In those scenarios, the HKF and its generalization are better choices for the optimization of filter gains.

4.4 Conclusion

LMMSE estimation with distributed smart sensors under perfect global model knowledge is solved by means of the distributed Kalman filter (DKF). Unfortunately, the estimate obtained at the fusion center is biased when the **measurement capacity** of the sensor network is not exactly known to all sensors. Indeed, the measurement capacity varies with nonlinear models, sensor failures, stochastic communication, etc. so that the areas of application of the DKF are confined to very basic scenarios with linear models. Consequently, the aim of this chapter was to reduce the dependence of the DKF on global model knowledge and to provide estimates for networks that are subject to communication losses and delays. To this end, generalizations of the DKF were proposed that successively reduced the degree of knowledge required for optimizing filter gains at the sensors.

In a first step, the measurement capacity was assumed known but due to unforeseeable sensor or communication failures, not all data necessary to reconstruct the LMMSE estimate was received at the fusion center. In this setup, **bias detection and correction** techniques prove effective in providing reliable estimates. Then, motivated by the sedate behavior of the measurement capacity in surveillance scenarios, filter gains were chosen according to a **hypothesis** instead of the true measurement capacity. The technique features a simple additive structure of estimates that enables the fusion of data already during communication. Extensions of the basic scheme to handle packet delays and losses were derived and the choice of the hypothesis was discussed. Although the technique already uses filter gains that are independent from global model knowledge, it still requires employing the same hypothesis at all sensors.

The most general form of hypothesizing distributed filtering was presented in Section 4.3 by deriving general optimization rules for the filtering and fusion of estimates. The idea is to use sensor-specific best guesses about the measurement capacity to calculate filter gains.

Indeed, the general form necessitates fusion techniques that were proposed based on the results of Chapter 3. In this context, bounding techniques for hypothesizing estimators were derived.

The next step would be to optimize filter gains in the presence of multiple fusion centers. However, as the filter processing at the sensors can satisfy the LMMSE criterion only for one fusion center, **conflicting measurement capacities** need to be reconciled. Depending on the weighting of fusion centers in the cost function, different filter gains are obtained. This can also be seen as preparatory work for an application of the decomposition techniques to decentralized estimation, where all nodes are construed as fusion centers and additionally to the aforementioned challenges, common prior information need to be considered.

Another open research question is to derive the optimal preprocessing when estimates are recursively comprised at the fusion center. Then, measurements are included infinitely often and the LMMSE filter processing depends on communication properties. Still, in contrast to the replacement strategy such an approach is directly applicable to decentralized estimation and reduces complexity as well as storage requirements at the fusion center.

CHAPTER 5

Conclusions and Future Research

Sensor network estimation is a large field of study with widespread areas of application. The underlying concept is to observe information at spatially distributed sensors, to exchange the potentially heterogeneous data that ideally provide different perspectives on the same phenomenon, and to merge them to draw new or better conclusions in control or decision problems.

For a mathematical treatment of the problem, information is abstractly modeled by means of random variables and the relation between measurements and an unknown state is assumed given. In this setup, the objective of discrete-time estimators is to optimize the processing in initialization, prediction, filter, and fusion steps. A meaningful criterion for evaluating the performance of estimators is the **mean squared error** (MSE), which is popular in the Bayesian framework and enables the derivation of strong theoretical statements.

Due to the sheer amount of sensor, processing, and communication types, it is almost impossible to provide general solutions that fit all estimation problems. In this thesis, sensor networks were classified into **centralized**, **distributed**, and **decentralized** processing schemes, and the focus was laid on linear systems and estimators. In the basic centralized setup, all observations are transmitted to a dedicated fusion center. The linear minimum mean squared error (LMMSE) estimator for this type of sensor network is the centralized Kalman filter. Distributed and decentralized schemes consider a (pre-)processing of measurements at the sensors. Therefore, communication resources can be exploited more efficiently than in the centralized scheme. However, the complexity of communication and processing is increased, which aggravates the analysis and derivation of proper estimators. Even though LMMSE filter and fusion techniques for distributed and decentralized estimation have been proposed in literature, they are hardly applicable to real world problems where the knowledge at sensors is restricted and the dependencies between estimates are concealed.

5.1 Summary of the Results

The focus of this thesis was laid on deriving adaptive and flexible estimators for distributed and decentralized estimation that function subject to limited sensor knowledge. The two key operations to be considered in this context are the **fusion**, i.e., the combination of estimates, and the **filtering**, i.e., the recursive processing of measurements.

Linear Fusion in Sensor Networks

In interconnected estimation systems, sensor information about the unknown state is gained from measurements and, in particular, by means of data exchanges. For a meaningful fusion and analysis of estimates, uncertainties of and dependencies between estimates must be exploited. Techniques for the reconstruction of these quantities and for the optimization of the fusion under imprecisely known uncertainties were derived in Chapter 3.

In the first instance, covariance estimation in linear systems was examined. To this end, the joint error density was estimated by means of samples. As prediction and filtering operations only necessitate updating local subsets of joint samples in linear systems, sample transformations can be decomposed into local operations that can be carried out at the sensors without access to transformations of remote nodes. Indeed, by combining sample sets from different nodes, e.g., in the fusion, cross-covariance matrices are obtained with standard covariance estimators. Notably, in the proposed scheme, sampling distribution and covariance estimator can be chosen freely. The decomposition of the central processing was derived and consistency of the **sample-based covariance estimator** was established. In order to enhance the precision of the scheme for finite sample sizes, different covariance estimators were discussed and the optimal sampling distribution in terms of the expected Frobenius norm was shown to be a modified Bernoulli distribution.

The second part of the chapter was concerned with the **partial reconstruction of cross-covariance matrices**. For linear models with independent measurement and process noise terms, (cross-)covariance matrices of estimates are specified by sums of linearly transformed noise covariances. A theoretic concept for the reconstruction of dependencies is to select and to recursively process square roots of noise terms individually at the sensors and to obtain (cross-)covariance matrices from their combination. By this means, sensors operate on locally known quantities without access to transformations of remote nodes. In order to limit the computational effort at the sensors, irrelevant terms can be aggregated. Then, the exchange of explicitly considered and aggregated noise terms enables the reconstruction of the joint covariance matrix as the sum of two terms, where cross-covariance matrices are exactly quantifiable for one term and the unknown cross-covariance matrices of the other term can be bounded.

For the obtained joint covariance matrix representation, fusion techniques were derived that **minimize the bound** and draw on well-established results for completely known and completely unknown correlations. As a matter of fact, the proposed theory is a generalization of the two concepts and permits the reconstruction and utilization of (cross-)covariance matrices of any precision subject to computational and communication effort. As part of the derivations, theoretical properties such as conservative bounds for the proposed covariance representation and convexity of a necessary fusion gain optimization were proven. These properties also apply to the underlying concepts and establish results for the fusion under unknown correlations not yet proven in literature. In particular, covariance intersection was shown to provide the smallest possible bound for the fusion of two estimates under completely unknown correlations.

Indeed, sample-based covariance estimation and covariance bounding have their own **assets and drawbacks**. For example, it is an inherent characteristic of bounds that the comparison of them – as required, for example, in decision problems – is a heuristic when no statement about the tightness of the bounds is provided. In addition, bounds tend to become conservative, i.e., the difference between bound and true covariance is large, when cross-covariance matrices are not partially exploited.

However, on the other hand, bounds give a reliable worst-case quality assessment of estimates that can be used in the calculation of filtering and fusion operations. In contrast to that, covariance estimates are subject to stochastic errors such that every evaluation based on the provided covariances is fraught with uncertainty, which renders the usage of estimated covariances difficult when stability properties or guaranteed quality are claimed.

Selecting the appropriate fusion technique depends on the application. Still, the precision and complexity of both approaches is adjustable such that customized solutions for many setups can be implemented. Considering the literature on decentralized estimation and, in particular, consensus [125] and diffusion approaches [27], the reconstruction and estimation of cross-covariance matrices based on locally processed variables can help to supplant fusion methods for time-invariant systems with theoretically justified techniques that minimize the mean squared error.

Distributed Filtering without Global Model Knowledge

The local incorporation of measurements into estimates largely determines the quality of estimates. Thanks to the Kalman filter, the associated theory for linear systems is well developed. Hence, the challenge of filtering in sensor networks lies in applying Kalman filter concepts to distributed and decentralized estimation schemes. For distributed estimation, a decomposition of the centralized Kalman filter that implements this idea is the distributed Kalman filter [59]. The decomposition scheme, however, is only applicable in very basic setups subject to deterministic communication. When models are time-variant or nonlinear, or when the communication is stochastic, the decomposition approach provides biased results at the fusion center.

Hence, Chapter 4 commenced with the derivation of additive and multiplicative debiasing techniques to **detect and correct the bias** of linear estimators in exceptional events like sensor failures. Indeed, in complex and uncertain environments, debiasing techniques alone do not suffice to provide precise estimates. The driving factor in minimizing the error at the fusion center was rather identified to be the optimization of filter gains according to a global measurement capacity, which determines the joint precision of all interconnected sensors. Consequently, the **hypothesizing distributed Kalman filter** was proposed that permits the optimization of filter gains according to a hypothesis about the global measurement capacity. Arising discrepancies between hypothesis and true measurement capacity are cushioned with the multiplicative debiasing technique. Moreover, methods to optimize hypotheses based on stochastic properties of the communication, i.e., in the presence of packet delays and losses, were discussed. As the proposed approach is a generalization of the Kalman filter decomposition scheme, LMMSE results are achieved for a proper chosen hypothesis. Otherwise, it was demonstrated by means of evaluations that hypothesizing approaches outperform classical sensor network estimators that utilize local Kalman filters.

Eventually, sensors that optimize the filter processing by means of **individual local hypotheses** were considered. This concept requires the least knowledge about the sensor network but necessitates a more elaborate fusion rule that was derived based on the bounding technique from Chapter 3. The bounding technique was also applied to variables obtained with the basic hypothesizing distributed Kalman filter to provide worst-case assessments of covariances in distributed estimation. Furthermore, a special type of hypothesizing filters was examined that is based on the scalar inflation of local measurement models, calculates unbiased estimates even without debiasing techniques, and provides the LMMSE results when all sensors are of the same type.

The hypothesizing filter optimization in sensor networks is a new concept that conceptually outperforms classical approaches that employ Kalman filters at the sensors. In particular, for interconnected systems with a large number of sensors, globally optimized filter gains enable a more efficient exploitation of measurement information that amounts to significant performance improvement. The basic idea of hypothesizing filters transfers to the estimator design in decentralized schemes so that future research in this area will systematically enhance state-of-the-art approaches.

5.2 Future Work

In this thesis, key problems of distributed and decentralized estimation were examined and solutions were proposed that are applicable to a range of real world problems. Indeed, it is the essence of research that any advance opens up new research questions. The following ideas portray only a few of many captivating topics that emerge from the thesis and that should be pursued to improve linear estimation theory for sensor networks.

Efficient Estimation Algorithms for Decentralized Processing Schemes

Considering the results on distributed estimation and reconstruction of cross-covariance matrices, the natural future research direction is to combine the gained knowledge about filter and fusion gain optimizations to derive efficient estimators for decentralized systems. For this purpose, it is necessary to optimize multiple estimates concurrently according to the sum or average of their mean squared errors, which **couples the optimization** of filter and fusion gains. As a matter of fact, the calculation of gains is competing in the sense that the optimized filtering of measurements according to the mean squared error of one node results in suboptimal processing with respect to the remaining nodes.

Moreover, in order to keep data amount and complexity manageable, estimates must be recursively fused in decentralized estimation. This is different from the proposed concept for distributed estimation where individual sensor packages were maintained and replaced at the fusion node. Consequently, LMMSE estimators for decentralized estimation depend on the communication between nodes.

Taking into account the complexity of the concurrent optimization of filter and fusion gains and the dependence on the stochastic communication, theoretic results on decentralized estimation will only be achieved in narrowly specified scenarios. It seems promising to devote special attention to systems with **time-invariant models and communication**. There, the optimization of filter and fusion processing according to an expected use of estimates in future time steps is possible. As the average mean squared error is not guaranteed to improve in every time step and, in particular, the LMMSE gains for one time step can yield suboptimal results when a horizon of two time steps is considered, the optimization of gains according to the steady state of the system is particularly interesting.

Even though achieving theoretically well-founded estimators for decentralized estimation in the LMMSE framework is still a long way away and will require considerable additional effort, taking into account the attention already turned on heuristically approaches for decentralized estimation, the interest and range of applications in research and practice makes it worthwhile.

Advances in Cross-covariance Matrix Reconstruction

The reconstruction of cross-covariance matrices constitutes an essential part in the fusion of estimates and in sensor network processing. Nevertheless, it is an auxiliary tool for state estimation and has not attracted much attention in literature yet.

From a mathematical perspective, the challenge of covariance estimation in linear systems boils down to **reconstructing a sum** of terms where the (transformed noise) terms are products of only locally known variables. In order to keep computational and communication effort limited, it is desired to combine these variables recursively at the sensors and still calculate the sum in the fusion. Now, different sum terms are relevant for cross-covariance matrices of different estimates. Thus, premised on the assumption that dependencies are recovered from recursively calculated variables, individual sum terms must be orthogonal to each other. Consequently, the ideas pursued in this thesis were either to process particular terms individually or to achieve the desired orthogonality asymptotically by generating noise samples independently from other noise terms.

Indeed, further improvements of the proposed techniques are possible. Some of the ideas, e.g., the smart selection of noise terms and the optimization of sampling distributions in multivariate systems, have already been discussed. It is, however, also conceivable to take a completely different approach in order to achieve the desired orthogonality. Considering that signals can be decomposed into complex sinus functions and the convolution of signals corresponds to an element-wise multiplication of the coefficients of these sinus functions, an idea is to assign each noise term a **different frequency** and to store sensor-specific information about the noise term on the corresponding frequency in the local signal. Then, estimator transformations are applied to local signals and the cross-covariance matrices are obtained by evaluating convoluted signals. Of course, the basic concept works only with scalar valued noise terms and only a finite number of frequencies can be used in practical applications. Still, the well-established signal processing theory can be used to optimize the processing of the variables and to find appropriate representations.

APPENDIX A

Linear Estimation Theory

Theorem A.1 (Orthogonality Principle) *Let $\underline{g}(\cdot)$ denote an arbitrary vector function of $\underline{\mathbf{z}}$, e.g., an estimator, and $\underline{\mathbf{x}}$ a random variable. Then, the error of the conditional expectation is orthogonal to $\underline{g}(\cdot)$, i.e.,*

$$\mathrm{E}_{\underline{\mathbf{x}},\underline{\mathbf{z}}}\{(\mathrm{E}_{\underline{\mathbf{x}}|\underline{\mathbf{z}}}\{\underline{\mathbf{x}}|\underline{\mathbf{z}}=\underline{z}\}-\underline{\mathbf{x}})^\top \underline{g}(\underline{\mathbf{z}})\} = 0 \ . \tag{A.1}$$

PROOF. Let $\underline{\widetilde{\mathbf{x}}}(\underline{\mathbf{z}})$ denote an arbitrary estimator. With $p(\underline{\mathbf{x}},\underline{\mathbf{z}}) = p(\underline{\mathbf{x}}|\underline{\mathbf{z}})p(\underline{\mathbf{z}})$, it follows

$$\mathrm{E}_{\underline{\mathbf{x}},\underline{\mathbf{z}}}\{(\underline{\widetilde{\mathbf{x}}}(\underline{\mathbf{z}})-\underline{\mathbf{x}})^\top \underline{g}(\underline{\mathbf{z}})\} = \mathrm{E}_{\underline{\mathbf{z}}}\{\mathrm{E}_{\underline{\mathbf{x}}|\underline{\mathbf{z}}}\{(\underline{\widetilde{\mathbf{x}}}(\underline{z})-\underline{\mathbf{x}})^\top \underline{g}(\underline{z})|\underline{\mathbf{z}}=\underline{z}\}\}$$

For given $\underline{z}$, $\underline{g}(\underline{z})$ and $\underline{\widetilde{\mathbf{x}}}(\underline{z})$ are constants. Thus, it holds

$$\begin{aligned}&\mathrm{E}_{\underline{\mathbf{z}}}\{\mathrm{E}_{\underline{\mathbf{x}}|\underline{\mathbf{z}}}\{(\underline{\widetilde{\mathbf{x}}}(\underline{z})-\underline{\mathbf{x}})^\top \underline{g}(\underline{z})|\underline{\mathbf{z}}=\underline{z}\}\} = \\ &\mathrm{E}_{\underline{\mathbf{z}}}\{(\underline{\widetilde{\mathbf{x}}}(\underline{z})-\mathrm{E}_{\underline{\mathbf{x}}|\underline{\mathbf{z}}}\{\underline{\mathbf{x}}|\underline{\mathbf{z}}=\underline{z}\})^\top \underline{g}(\underline{z})\} \ .\end{aligned}$$

This term vanishes for $\underline{\widetilde{\mathbf{x}}}(\underline{\mathbf{z}}) = \mathrm{E}_{\underline{\mathbf{x}}|\underline{\mathbf{z}}}\{\underline{\mathbf{x}}|\underline{\mathbf{z}}=\underline{z}\}$, which concludes the proof. □

Proof of Theorem 2.1 Let $\underline{\widetilde{\mathbf{x}}}(\underline{\mathbf{z}})$ denote an arbitrary estimator and let $\underline{\hat{\mathbf{x}}}(\underline{\mathbf{z}}) = \mathrm{E}\{\underline{\mathbf{x}}|\underline{\mathbf{z}}=\underline{z}\}$ denote the conditional expectation. It holds

$$\begin{aligned}\mathrm{E}\{\ell_2(\underline{\widetilde{\mathbf{x}}}(\underline{\mathbf{z}}),\underline{\mathbf{x}})\} &= \mathrm{E}\{\ell_2(\underline{\widetilde{\mathbf{x}}}(\underline{\mathbf{z}})-\underline{\hat{\mathbf{x}}}(\underline{\mathbf{z}}),\underline{\mathbf{x}}-\underline{\hat{\mathbf{x}}}(\underline{\mathbf{z}}))\} \\ &= \mathrm{E}\{\ell_2(\underline{\widetilde{\mathbf{x}}}(\underline{\mathbf{z}}),\underline{\hat{\mathbf{x}}}(\underline{\mathbf{z}}))\} + \mathrm{E}\{\ell_2(\underline{\mathbf{x}},\underline{\hat{\mathbf{x}}}(\underline{\mathbf{z}}))\} + \\ &\quad 2\cdot \mathrm{E}\{(\underline{\widetilde{\mathbf{x}}}(\underline{\mathbf{z}})-\underline{\hat{\mathbf{x}}}(\underline{\mathbf{z}}))^\top(\underline{\mathbf{x}}-\underline{\hat{\mathbf{x}}}(\underline{\mathbf{z}}))\} \ .\end{aligned}$$

Splitting up the third term yields

$$\begin{aligned} \mathrm{E}\{(\underline{\tilde{\mathbf{x}}}(\underline{\mathbf{z}}) - \underline{\hat{\mathbf{x}}}(\underline{\mathbf{z}}))^\top (\underline{\mathbf{x}} - \underline{\hat{\mathbf{x}}}(\underline{\mathbf{z}}))\} = &\mathrm{E}\{\underline{\tilde{\mathbf{x}}}(\underline{\mathbf{z}})^\top (\underline{\mathbf{x}} - \underline{\hat{\mathbf{x}}}(\underline{\mathbf{z}}))^\top\} - \\ &\mathrm{E}\{\underline{\hat{\mathbf{x}}}(\underline{\mathbf{z}}))(\underline{\mathbf{x}} - \underline{\hat{\mathbf{x}}}(\underline{\mathbf{z}}))^\top\} \end{aligned}$$

which vanishes according to Theorem A.1. As

$$\mathrm{E}\{\ell_2(\underline{\tilde{\mathbf{x}}}(\underline{\mathbf{z}}), \underline{\hat{\mathbf{x}}}(\underline{\mathbf{z}}))\} \geq 0 \; ,$$

it holds

$$\begin{aligned} \mathrm{E}\{\ell_2(\underline{\tilde{\mathbf{x}}}(\underline{\mathbf{z}}), \underline{\mathbf{x}})\} &= \mathrm{E}\{\ell_2(\underline{\tilde{\mathbf{x}}}(\underline{\mathbf{z}}), \underline{\hat{\mathbf{x}}}(\underline{\mathbf{z}}))\} + \mathrm{E}\{\ell_2(\underline{\mathbf{x}}, \underline{\hat{\mathbf{x}}}(\underline{\mathbf{z}}))\} \\ &\geq \mathrm{E}\{\ell_2(\underline{\mathbf{x}}, \underline{\hat{\mathbf{x}}}(\underline{\mathbf{z}}))\} \; , \end{aligned}$$

which concludes the proof. □

APPENDIX B

Sample-based Covariance Reconstruction

Proof of Theorem 3.1 The statement is proven inductively. For this purpose, let $(\hat{\underline{\mathbf{x}}}_0, \Phi_0)$ denote a valid sample representation for the state $\underline{\mathbf{x}}_0$. It is shown for all operations of Algorithm 3.2 that the sample representation remains valid. As process noise is generated with synchronized samplers and measurement noise with independent ones, the generation implements the policy described in Definition 3.3. In order to prove correctness of the prediction operation, $(\mathbf{A}\hat{\underline{\mathbf{x}}}, \mathbf{A}\Phi + \Phi^w)$ must be a valid sample representation of $\mathbf{A}\underline{\mathbf{x}} + \underline{\mathbf{w}}$. Consistent covariance estimators inherit the bilinearity of covariances for $d \to \infty$. Thus, $\lim_{d\to\infty} \hat{\mathbf{C}}(\mathbf{T}\Phi + \Phi^w, \mathbf{T}\Phi + \Phi^w)$ equals

$$\lim_{d\to\infty} (\mathbf{T}\hat{\mathbf{C}}(\Phi)(\mathbf{T})^\top + \hat{\mathbf{C}}(\Phi^w, \Phi^w) + \hat{\mathbf{C}}(\mathbf{T}\Phi, \Phi^w) + \hat{\mathbf{C}}(\Phi^w, \mathbf{T}\Phi)) \ .$$

Let $\mathbf{C}$ denote the true covariance of $\hat{\underline{\mathbf{x}}}$. As Φ^w is sampled independently of Φ, the terms $\hat{\mathbf{C}}(\mathbf{T}\Phi, \Phi^w)$ vanish, yielding

$$\begin{aligned}\lim_{d\to\infty} \hat{\mathbf{C}}(\mathbf{T}\Phi + \Phi^w, \mathbf{T}\Phi + \Phi^w) = &\lim_{d\to\infty} (\mathbf{T}\hat{\mathbf{C}}(\Phi)(\mathbf{T})^\top + \hat{\mathbf{C}}(\Phi^w, \Phi^w)) \\ =&\mathbf{T}\mathbf{C}(\mathbf{T})^\top + \mathbf{Q} \ ,\end{aligned}$$

which corresponds to the true covariance of $\underline{\mathbf{e}} = \mathbf{A}\hat{\underline{\mathbf{x}}} - \mathbf{A}\underline{\mathbf{x}} - \underline{\mathbf{w}}$. Analogous considerations lead to

$$\lim_{d\to\infty} \hat{\mathbf{C}}(\mathbf{L}\Phi + \mathbf{K}\Phi^v, \mathbf{L}\Phi + \mathbf{K}\Phi^v) = \mathbf{L}\mathbf{C}(\mathbf{L})^\top + \mathbf{K}\mathbf{R}(\mathbf{K})^\top \ ,$$

i.e., the covariance of $\underline{\mathbf{e}} = \mathbf{L}\hat{\underline{\mathbf{x}}} + \mathbf{K}\underline{\mathbf{z}} - \underline{\mathbf{x}} = (\mathbf{I} - \mathbf{K}\mathbf{H})(\hat{\underline{\mathbf{x}}} - \underline{\mathbf{x}}) + \mathbf{K}\underline{\mathbf{v}}$. Now, let $(\hat{\underline{\mathbf{x}}}^s, \Phi^s)$ denote valid sample representations. As comprised noise terms are sampled according to valid sampling policies and all transformations are linear, it is known from (3.7) that the samples are distributed according to error processes $\Phi^s \sim \sum_{i\in\mathcal{I}^s} \mathbf{T}_i^s \underline{\boldsymbol{\psi}}_i$. It remains to show that $\sum_{s\in\mathcal{S}} \mathbf{F}^s\Phi^s \sim \sum_{s\in\mathcal{S}} \mathbf{F}^s\hat{\underline{\mathbf{x}}}^s - \underline{\mathbf{x}} = \underline{\mathbf{e}}$. From the asymptotic bilinearity of consistent covariance estimators it follows

$$\begin{aligned}\lim_{d\to\infty} \hat{\mathbf{C}}\left(\sum_{s\in\mathcal{S}} \mathbf{F}^s\Phi^s\right) &= \lim_{d\to\infty} \sum_{s,\tilde{s}\in\mathcal{S}} \mathbf{F}^s\hat{\mathbf{C}}(\Phi^s, \Phi^{\tilde{s}})(\mathbf{F}^{\tilde{s}})^\top \\ &= \lim_{d\to\infty} \sum_{s,\tilde{s}\in\mathcal{S}} \mathbf{F}^s \sum_{i,\tilde{i}\in\mathcal{I}^s\cup\mathcal{I}^{\tilde{s}}} \mathbf{T}_i^s\hat{\mathbf{C}}(\Phi_i^s, \Phi_{\tilde{i}}^{\tilde{s}})(\mathbf{T}_{\tilde{i}}^{\tilde{s}})^\top(\mathbf{F}^{\tilde{s}})^\top ,\end{aligned}$$

where $\lim_{d\to\infty} \hat{\mathbf{C}}(\Phi_i^s, \Phi_{\tilde{i}}^{\tilde{s}}) = \mathbf{0}$ for $i \neq \tilde{i}$ and $\lim_{d\to\infty} \hat{\mathbf{C}}(\Phi_i^s, \Phi_{\tilde{i}}^{\tilde{s}}) = \mathbf{C}_i^\nu$ for $i = \tilde{i}$ due to Definition 3.5. Hence,

$$\begin{aligned}\lim_{d\to\infty} \hat{\mathbf{C}}\left(\sum_{s\in\mathcal{S}} \mathbf{F}^s\Phi^s\right) &= \sum_{s,\tilde{s}\in\mathcal{S}} \mathbf{F}^s \sum_{i\in\mathcal{I}^s\cap\mathcal{I}^{\tilde{s}}} \mathbf{T}_i^s \lim_{d\to\infty} \hat{\mathbf{C}}(\Phi_i^s, \Phi_i^{\tilde{s}})(\mathbf{T}_i^{\tilde{s}})^\top(\mathbf{F}^{\tilde{s}})^\top \\ &= \sum_{s,\tilde{s}\in\mathcal{S}} \mathbf{F}^s \sum_{i\in\mathcal{I}^s\cap\mathcal{I}^{\tilde{s}}} \mathbf{T}_i^s\mathbf{C}_i^\nu(\mathbf{T}_i^{\tilde{s}})^\top(\mathbf{F}^{\tilde{s}})^\top ,\end{aligned}$$

which corresponds to the true error of $\sum_{s\in\mathcal{S}} \mathbf{F}^s\hat{\underline{\mathbf{x}}}^s$ according to Theorem 3.17 with (3.35). □

Proof of Lemma 3.2 Consider the distribution $\tilde{\boldsymbol{\phi}} = (\boldsymbol{\phi}^{\nu_1} - \mathrm{E}\{\boldsymbol{\phi}^{\nu_1}\})(\boldsymbol{\phi}^{\nu_2} - \mathrm{E}\{\boldsymbol{\phi}^{\nu_2}\})$. The natural variance estimator $\hat{\mathbf{v}}$ can be construed as the mean of samples $\tilde{\phi}_1, \ldots, \tilde{\phi}_d$ from the distribution $\tilde{\boldsymbol{\phi}}$. Therefore, it holds

$$\begin{aligned}\mathrm{E}\{\hat{\mathbf{v}}(\Phi^{\nu_1}, \Phi^{\nu_2})\} &= \mathrm{E}\left\{\frac{1}{d}\sum_{i=1}^{d}(\phi_i^{\nu_1} - \mathrm{E}\{\boldsymbol{\phi}^{\nu_1}\})(\phi_i^{\nu_2} - \mathrm{E}\{\boldsymbol{\phi}^{\nu_2}\})\right\} \\ &= \frac{1}{d}\sum_{i=1}^{d} \mathrm{E}\{\tilde{\phi}_i\} \\ &= \mathrm{E}\{\tilde{\boldsymbol{\phi}}\} .\end{aligned}$$

The desired variance is given by

$$\mathbf{var}\left(\hat{\mathbf{v}}(\Phi^{\nu_1},\Phi^{\nu_2})\right) = \mathrm{E}\{(\hat{\mathbf{v}}(\Phi^{\nu_1},\Phi^{\nu_2}))^2\} - \mathrm{E}\{\hat{\mathbf{v}}(\Phi^{\nu_1},\Phi^{\nu_2})\}^2 \ ,$$

where

$$\begin{aligned}\mathrm{E}\{(\hat{\mathbf{v}}(\Phi^{\nu_1},\Phi^{\nu_2}))^2\} =& \mathrm{E}\left\{\Big(\frac{1}{d}\sum_{i=1}^{d}\tilde{\phi}_i\Big)^2\right\}\\ =& \frac{1}{d^2}\sum_{i,\tilde{i}=1}^{d}\mathrm{E}\{\tilde{\phi}_i\tilde{\phi}_{\tilde{i}}\} \ .\end{aligned}$$

From the independence of samples $\tilde{\phi}$, it follows $\mathrm{E}\{\tilde{\phi}_i\tilde{\phi}_{\tilde{i}}\} = \mathrm{E}\{\tilde{\boldsymbol{\phi}}\}^2$ for $i \neq \tilde{i}$. Thus, it holds

$$\mathrm{E}\{(\hat{\mathbf{v}}(\Phi^{\nu_1},\Phi^{\nu_2}))^2\} = \frac{1}{d}\,\mathrm{E}\{\tilde{\boldsymbol{\phi}}^2\} + \frac{d(d-1)}{d^2}\,\mathrm{E}\{\tilde{\boldsymbol{\phi}}\}^2 \ .$$

The desired equation is obtained with

$$\mathbf{var}\left(\hat{\mathbf{v}}(\Phi^{\nu_1},\Phi^{\nu_2})\right) = \frac{1}{d}\,\mathrm{E}\{\tilde{\boldsymbol{\phi}}^2\} + \frac{d-1}{d}\,\mathrm{E}\{\tilde{\boldsymbol{\phi}}\}^2 - \mathrm{E}\{\tilde{\boldsymbol{\phi}}\}^2 \ ,$$

where $\mathrm{E}\{\tilde{\boldsymbol{\phi}}\} = \mathbf{var}\left(\boldsymbol{\phi}^{\nu_1},\boldsymbol{\phi}^{\nu_2}\right)$. □

Proof of Theorem 3.4 According to Lemma 3.2 it holds

$$\mathbf{var}\left(\hat{\mathbf{v}}(\Phi)\right) = \frac{1}{d}\left(\mathrm{E}\{(\boldsymbol{\phi} - \mathrm{E}\{\boldsymbol{\phi}\})^4\} - \left(\mathbf{var}\left(\boldsymbol{\phi}\right)\right)^2\right) \ .$$

Considering the sum representation of the error process (3.7), $\mathrm{E}\{(\boldsymbol{\phi} - \mathrm{E}\{\boldsymbol{\phi}\})^4\}$ equals

$$\sum_{j_1,j_2,j_3,j_4\in\mathcal{I}} \mathbf{T}_{j_1}\cdots\mathbf{T}_{j_4}\,\mathrm{E}\{(\boldsymbol{\phi}^{\nu}_{j_1} - \mathrm{E}\{\boldsymbol{\phi}^{\nu}_{j_1}\})\cdots(\boldsymbol{\phi}^{\nu}_{j_4} - \mathrm{E}\{\boldsymbol{\phi}^{\nu}_{j_4}\})\} \ .$$

From the independence of noise terms, it follows that the expectation term can be separated into terms with different indices. Terms with a single occurrence of an index, e.g., $j_1 \notin \{j_2, j_3, j_4\}$, vanish as

$\mathrm{E}\{\phi^\nu - \mathrm{E}\{\phi^\nu\}\} = 0$. Therefore, only terms with (two times) two and four equal indices must be considered. The combinations with two different indices appear $\binom{4}{2} = 6$ times, amounting to the following sum

$$3 \cdot \sum_{j_1 \neq j_2 \in \mathcal{I}} (\mathbf{T}_{j_1})^2 (\mathbf{T}_{j_2})^2 \mathbf{var}\left(\phi^\nu_{j_1}\right) \mathbf{var}\left(\phi^\nu_{j_2}\right) .$$

The fourth order terms emerge only once per noise term, where $\mathrm{E}\{(\phi^\nu - \mathrm{E}\{\phi^\nu\})^4\} = (\mathbf{var}(\phi^\nu))^2$ for the modified Bernoulli distribution. Hence, the first part of the expected Frobenius norm is given by

$$\begin{aligned} \mathrm{E}\{(\phi - \mathrm{E}\{\phi\})^4\} =& 3 \cdot \sum_{j_1 \neq j_2 \in \mathcal{I}} (\mathbf{T}_{j_1})^2 (\mathbf{T}_{j_2})^2 \mathbf{var}\left(\phi^\nu_{j_1}\right) \mathbf{var}\left(\phi^\nu_{j_2}\right) + \\ & \sum_{j \in \mathcal{I}} (\mathbf{T}_j)^4 \left(\mathbf{var}(\phi^\nu)\right)^2 . \end{aligned}$$

Considering the variance of the estimate, which is given by

$$\begin{aligned} (\mathbf{var}(\phi))^2 =& \Big(\sum_{j \in \mathcal{I}} (\mathbf{T}_j)^2 \mathbf{var}(\phi^\nu) \Big)^2 \\ =& \sum_{j_1, j_2 \in \mathcal{I}} (\mathbf{T}_{j_1})^2 (\mathbf{T}_{j_2})^2 \mathbf{var}\left(\phi^\nu_{j_1}\right) \mathbf{var}\left(\phi^\nu_{j_2}\right) , \end{aligned}$$

the variance specified in Lemma 3.2 simplifies to

$$\begin{aligned} \mathbf{var}(\hat{\mathbf{v}}(\Phi)) =& \frac{2}{d} \sum_{j_1 \neq j_2 \in \mathcal{I}} (\mathbf{T}_{j_1})^2 (\mathbf{T}_{j_2})^2 \mathbf{var}\left(\phi^\nu_{j_1}\right) \mathbf{var}\left(\phi^\nu_{j_2}\right) \\ =& \frac{2}{d} \Big((\mathbf{var}(\mathbf{C}))^2 - \sum_{j \in \mathcal{I}} (\mathbf{T}_j)^4 (\mathbf{var}\left(\phi^\nu_j\right))^2 \Big) , \end{aligned}$$

which concludes the proof. □

Proof of Lemma 3.5 It holds

$$\mathrm{tr}\left\{\mathbf{Cov}\left(\hat{\mathbf{C}}\right)\right\} = \mathrm{tr}\left\{\mathrm{E}\{(\hat{\mathbf{E}})^2\}\right\} - \mathrm{tr}\left\{\mathbf{E}^2\right\} , \tag{B.1}$$

where $\hat{\mathbf{E}} = \frac{1}{d}\sum_{i=1}^{d}\left(\underline{\phi}_i - \mathrm{E}\{\underline{\boldsymbol{\phi}}\}\right)^2$. Therefore, it holds

$$\mathrm{E}\{\hat{\mathbf{E}}(\hat{\mathbf{E}})^\top\} = \frac{1}{d^2}\sum_{i,\tilde{i}=1}^{d} \mathrm{E}\{(\underline{\phi}_i - \mathrm{E}\{\underline{\boldsymbol{\phi}}\})^2(\underline{\phi}_{\tilde{i}} - \mathrm{E}\{\underline{\boldsymbol{\phi}}\})^2\} \ .$$

For independently generated samples $i \neq \tilde{i}$, the sum terms are given by

$$\begin{aligned}\mathrm{E}\{(\underline{\phi}_i{-}\mathrm{E}\{\underline{\boldsymbol{\phi}}\})^2(\underline{\phi}_{\tilde{i}}{-}\mathrm{E}\{\underline{\boldsymbol{\phi}}\})^2\} &= \mathrm{E}\{(\underline{\phi}_i{-}\mathrm{E}\{\underline{\boldsymbol{\phi}}\})^2\}\,\mathrm{E}\{(\underline{\phi}_{\tilde{i}}{-}\mathrm{E}\{\underline{\boldsymbol{\phi}}\})^2\} \\ &= \mathbf{E}^2 \ .\end{aligned}$$

If $i = \tilde{i}$, $\mathrm{E}\{(\underline{\phi}_i - \mathrm{E}\{\underline{\boldsymbol{\phi}}\})^2(\underline{\phi}_{\tilde{i}} - \mathrm{E}\{\underline{\boldsymbol{\phi}}\})^2\}$ is obtained by

$$\mathrm{E}\left\{\sum_{\rho=1}^{n_x}([\underline{\phi} - \mathrm{E}\{\underline{\boldsymbol{\phi}}\}]_\rho)^2 \begin{pmatrix} \epsilon_1^2 & \cdots & \epsilon_1\epsilon_{n_x} \\ \vdots & \ddots & \vdots \\ \epsilon_1\epsilon_{n_x} & \cdots & \epsilon_{n_x}^2 \end{pmatrix}\right\} ,$$

with $\epsilon_\rho = [\underline{\phi} - \mathrm{E}\{\underline{\boldsymbol{\phi}}\}]_\rho$, $\rho \in \{1, \ldots, n_x\}$. Therefore, $\mathrm{tr}\left\{\mathrm{E}\{\hat{\mathbf{E}}(\hat{\mathbf{E}})^\top\}\right\}$ equals

$$\frac{1}{d}\sum_{\rho,\tilde{\rho}=1}^{n_x} \mathrm{E}\{([\underline{\boldsymbol{\phi}} - \mathrm{E}\{\underline{\boldsymbol{\phi}}\}]_\rho)^2([\underline{\boldsymbol{\phi}} - \mathrm{E}\{\underline{\boldsymbol{\phi}}\}]_{\tilde{\rho}})^2\} + \frac{d-1}{d}\,\mathrm{tr}\left\{\mathbf{E}^2\right\} \ .$$

The claim follows with (B.1). □

Proof of Theorem 3.6 From Lemma 3.5 it is known that $(\underline{\mathbf{1}})^\top\breve{\mathbf{C}}\underline{\mathbf{1}}$ is to be minimized. Without loss of generality, let $\mathrm{E}\{\underline{\boldsymbol{\phi}}\} = \underline{0}$. Then,

$$(\underline{\mathbf{1}})^\top\breve{\mathbf{C}}\underline{\mathbf{1}} = (\underline{\mathbf{1}})^\top\,\mathrm{E}\{(\tilde{\underline{\boldsymbol{\phi}}})^2\}\underline{\mathbf{1}} \ ,$$

where

$$_\rho = ([\underline{\boldsymbol{\phi}}]_\rho)^2 \ , \rho \in \{1, \ldots, n_x\} \ .$$

Let $\tilde{\mathbf{C}}$ denote the positive semi-definite covariance of $\underline{\tilde{\phi}}$ with $\tilde{\mathbf{C}} = \mathrm{E}\{\underline{\tilde{\phi}}^2\} - \mathrm{E}\{\underline{\tilde{\phi}}\}^2$. It holds $\mathrm{E}\{\underline{\tilde{\phi}}\} = \mathrm{diag}\,(\mathbf{E})$, and thus, $\mathrm{E}\{(\underline{\tilde{\phi}})^2\} = \tilde{\mathbf{C}} + \mathbf{E}^2$. Therefore, $(\underline{1})^\top \mathrm{E}\{(\underline{\tilde{\phi}})^2\}\underline{1}$ is minimized when $(\underline{1})^\top \tilde{\mathbf{C}}\underline{1} = 0$, i.e., when $\underline{\tilde{\phi}}$ is deterministic.

From Theorem 3.4 it is known that $([\underline{\phi}]_\rho)^2$, $\rho \in \{1, \ldots, n_x\}$ is deterministic for Bernoulli distributed samples, i.e., the diagonal of $\tilde{\mathbf{C}}$ is $\underline{0}$, which implies that the entire covariance is $\mathbf{0}$. The claim follows as the multivariate modified Bernoulli distribution satisfies $\mathrm{E}\{(\underline{\phi} - \mathrm{E}\{\underline{\phi}\})^2\} = \mathbf{E}$.

□

APPENDIX C

Convexity of Generalized Covariance Intersection

Proof of Theorem 3.15 Let the variables be defined as in Theorem 3.14 and let $\bar{\mathbf{P}}_b = (\mathbf{\Omega})^{-1}\bar{\mathbf{C}}_b$ denote the joint space segmentation into weighting terms and corresponding (uninflated) covariances. For weights $\{\omega^s\}_{s\in\mathcal{S}}$, the derivative of the fused covariance bound is obtained with basic matrix calculus, e.g., with [130], as

$$\frac{\partial \mathbf{P}}{\partial \omega^s} = -\,\mathbf{P}\underline{\mathbf{1}}^\top \frac{\partial (\bar{\mathbf{P}})^{-1}}{\partial \omega^s}\underline{\mathbf{1}}\mathbf{P}\ ,$$

where

$$\frac{\partial (\bar{\mathbf{P}})^{-1}}{\partial \omega^s} = -(\bar{\mathbf{P}})^{-1}\frac{\partial (\mathbf{\Omega})^{-1}}{\partial \omega^s}\bar{\mathbf{C}}_b(\bar{\mathbf{P}})^{-1}\ .$$

Let $\bar{\mathbf{C}}_b^s$ denote the joint space matrix that has the same entries as $\bar{\mathbf{C}}_b$ in the s-th diagonal block and is zero otherwise. Then,

$$\frac{\partial (\mathbf{\Omega})^{-1}}{\partial \omega^s}\bar{\mathbf{C}}_b = -\frac{1}{(\omega^s)^2}\bar{\mathbf{C}}_b^s$$

and thus, the derivative of the bound of the fused estimate is given by the negative definite matrix

$$\frac{\partial \mathbf{P}}{\partial \omega^s} = -\frac{1}{(\omega^s)^2}\left(\mathbf{P}\underline{\mathbf{1}}^\top(\bar{\mathbf{P}})^{-1}\right)\bar{\mathbf{C}}_b^s\left(\mathbf{P}\underline{\mathbf{1}}^\top(\bar{\mathbf{P}})^{-1}\right)^\top\ .$$

Let $\mathbf{G}^s$ denote an auxiliary variable with $\mathbf{G}^s = (\omega^s)^{-2}\mathbf{P}\underline{\mathbf{1}}^\top(\bar{\mathbf{P}})^{-1}$. With

$$\frac{\partial\left(\mathbf{P}\underline{\mathbf{1}}^\top(\bar{\mathbf{P}})^{-1}\right)}{\partial\omega^s} = \frac{\partial\mathbf{P}}{\partial\omega^s}\underline{\mathbf{1}}^\top(\bar{\mathbf{P}})^{-1} + \mathbf{P}\underline{\mathbf{1}}^\top\frac{\partial(\bar{\mathbf{P}})^{-1}}{\partial\omega^s}$$
$$= \mathbf{G}^s\bar{\mathbf{C}}_b^s\left((\bar{\mathbf{P}})^{-1} - (\bar{\mathbf{P}})^{-1}\underline{\mathbf{1}}\mathbf{P}\underline{\mathbf{1}}^\top(\bar{\mathbf{P}})^{-1}\right),$$

and the symmetry of $\frac{\partial\mathbf{P}}{\partial\omega^s}$, the subsequent application of the chain rule leads to second derivatives

$$2\cdot\mathbf{G}^s\left(\omega^s\bar{\mathbf{C}}_b^s - \bar{\mathbf{C}}_b^s(\bar{\mathbf{P}})^{-1}\bar{\mathbf{C}}_b^s + \bar{\mathbf{C}}_b^s(\bar{\mathbf{P}})^{-1}\underline{\mathbf{1}}\mathbf{P}\underline{\mathbf{1}}^\top(\bar{\mathbf{P}})^{-1}\bar{\mathbf{C}}_b^s\right)(\mathbf{G}^s)^\top$$

for $\frac{\partial^2\mathbf{P}}{\partial(\omega^s)^2}$ and

$$\mathbf{G}^s\left(\mathbf{\Phi}^s(\bar{\mathbf{P}})^{-1}\underline{\mathbf{1}}\mathbf{P}\underline{\mathbf{1}}^\top(\bar{\mathbf{P}})^{-1}\mathbf{\Phi}^{\tilde{s}} - \mathbf{\Phi}^s(\bar{\mathbf{P}})^{-1}\mathbf{\Phi}^{\tilde{s}}\right)(\mathbf{G}^{\tilde{s}})^\top + (\,\cdot\,)^\top$$

for $\frac{\partial^2\mathbf{P}}{\partial\omega^s\partial\omega^{\tilde{s}}}$. Therefore, the Hessian matrix is given as the sum $\mathcal{H} = \tilde{\mathcal{H}} + (\tilde{\mathcal{H}})^\top$, where $\tilde{\mathcal{H}}$ equals

$$\underline{\mathbf{G}\mathbf{\Phi}}(\bar{\mathbf{P}})^{-1}\underline{\mathbf{1}}\mathbf{P}\underline{\mathbf{1}}^\top(\bar{\mathbf{P}})^{-1}(\underline{\mathbf{G}\mathbf{\Phi}})^\top + \underline{\mathbf{G}}\mathbf{\Omega}\mathbf{\Phi}(\underline{\mathbf{G}})^\top - \underline{\mathbf{G}\mathbf{\Phi}}(\bar{\mathbf{P}})^{-1}(\underline{\mathbf{G}\mathbf{\Phi}})^\top,$$

and

$$\underline{\mathbf{G}\mathbf{\Phi}} = \begin{pmatrix}\mathbf{G}^s\mathbf{\Phi}^s\\ \vdots\\ \mathbf{G}^{S_S}\mathbf{\Phi}^{S_S}\end{pmatrix} \text{ and } \mathbf{\Omega}\mathbf{\Phi} = \begin{pmatrix}\omega^s\mathbf{\Phi}^s & \dots & \mathbf{0}\\ \vdots & \ddots & \vdots\\ \mathbf{0} & \dots & \omega^{S_S}\mathbf{\Phi}^{S_S}\end{pmatrix}.$$

For positive semi-definite $\bar{\mathbf{C}}$ it holds

$$(\bar{\mathbf{P}})^{-1} = (\bar{\mathbf{C}} + \bar{\mathbf{P}}_b)^{-1} \leq (\bar{\mathbf{P}}_b)^{-1}$$

and therefore,

$$\underline{\mathbf{G}\mathbf{\Phi}}(\bar{\mathbf{P}})^{-1}(\underline{\mathbf{G}\mathbf{\Phi}})^\top \leq \underline{\mathbf{G}\mathbf{\Phi}}(\bar{\mathbf{P}}_b)^{-1}(\underline{\mathbf{G}\mathbf{\Phi}})^\top.$$

With $\mathbf{\Phi}^s\left(\bar{\mathbf{P}}\right)^{-1}\mathbf{\Phi}^{\tilde{s}} = \mathbf{0}$ for $s \neq \tilde{s}$ and $\mathbf{\Phi}^s\left(\bar{\mathbf{P}}\right)^{-1}\mathbf{\Phi}^{\tilde{s}} = \omega^s\mathbf{\Phi}^s$ for $s = \tilde{s}$, it holds

$$\underline{\mathbf{G}}\mathbf{\Omega}\mathbf{\Phi}(\underline{\mathbf{G}})^\top - \underline{\underline{\mathbf{G}\mathbf{\Phi}}}\left(\bar{\mathbf{P}}\right)^{-1}(\underline{\underline{\mathbf{G}\mathbf{\Phi}}})^\top \geq \mathbf{0} \,,$$

which proves that $\tilde{\mathcal{H}} \geq \mathbf{0}$. Therefore, the Hessian matrix is positive semi-definite. As

$$\frac{\partial^2 \operatorname{tr}\{\mathbf{P}\}}{\partial\omega^s\partial\omega^{\tilde{s}}} = \operatorname{tr}\left\{\frac{\partial^2\mathbf{P}}{\partial\omega^s\partial\omega^{\tilde{s}}}\right\} \,,$$

convexity follows when the trace joint covariance matrix

$$\bar{\mathbf{C}}^{tr} = \begin{pmatrix} \operatorname{tr}\left\{\mathbf{C}^{s_1}\right\} & \dots & \operatorname{tr}\left\{\mathbf{C}^{s_1 s_S}\right\} \\ \vdots & \ddots & \vdots \\ \operatorname{tr}\left\{\mathbf{C}^{s_S s_1}\right\} & \dots & \operatorname{tr}\left\{\mathbf{C}^{S}\right\} \end{pmatrix}$$

is positive semi-definite. Let $\underline{\sigma}^s$ denote vectors with standard deviations for covariances $\mathbf{C}^s$, and let $\mathbf{R}^{s\tilde{s}}$ denote the respective correlation matrices such that $\mathbf{C}^{s\tilde{s}} = \operatorname{diag}(\underline{\sigma}^s)\mathbf{R}^{s\tilde{s}}\operatorname{diag}\left(\underline{\sigma}^{\tilde{s}}\right)$. It holds $\operatorname{tr}\left\{\mathbf{C}^{s\tilde{s}}\right\} = (\underline{\sigma}^s)^\top \operatorname{diag}\left(\underline{\rho}^{s\tilde{s}}\right)\underline{\sigma}^{\tilde{s}}$, where $\underline{\rho}^{s\tilde{s}} = \operatorname{diag}\left(\mathbf{R}^{s\tilde{s}}\right)$ denotes the diagonal of the correlation matrices. Hence, the trace joint covariance matrix is given by

$$\bar{\mathbf{C}}^{tr} = (\mathbf{\Sigma})^\top \underbrace{\begin{pmatrix} \operatorname{diag}\left(\underline{\rho}^{s_1}\right) & \dots & \operatorname{diag}\left(\underline{\rho}^{s_1 s_S}\right) \\ \vdots & \ddots & \vdots \\ \operatorname{diag}\left(\underline{\rho}^{s_1 s_S}\right) & \dots & \operatorname{diag}\left(\underline{\rho}^{s_S}\right) \end{pmatrix}}_{\mathbf{R}_{diag}} \mathbf{\Sigma} \,,$$

where

$$\mathbf{\Sigma} = \begin{pmatrix} \underline{\sigma}^1 & \dots & \underline{0} \\ \vdots & \ddots & \vdots \\ \underline{0} & \dots & \underline{\sigma}^{s_S} \end{pmatrix} .$$

Therefore, $\bar{\mathbf{C}}^{tr} \geq \mathbf{0}$, i.e., the claim is satisfied, if the joint correlation matrix $\mathbf{R}_{diag}$ is positive semi-definite[1]. Now consider the permutation of elements π such that $\underline{\sigma}^{\pi_i} = ([\underline{\sigma}^1]_i, \ldots, [\underline{\sigma}^S]_i)^\top$ and let $\mathbf{R}^{\pi}_{diag}$ denote the respective joint correlation matrix. Then, $\mathbf{R}^{\pi}_{diag}$ is a block diagonal matrix with blocks

$$\begin{pmatrix} [\mathbf{R}^{s_1}]_{ii} & \cdots & [\mathbf{R}^{s_1 s_S}]_{ii} \\ \vdots & \ddots & \vdots \\ [\mathbf{R}^{s_S s_1}]_{ii} & \cdots & [\mathbf{R}^{s_S}]_{ii} \end{pmatrix} .$$

The blocks are correlation matrices of permuted random variables. Therefore, they are positive semi-definite. Hence, the matrix $\mathbf{R}^{\pi}_{diag}$ and its permuted counterpart $\mathbf{R}_{diag}$ are positive semi-definite, which concludes the proof. □

[1] The removal of entries from a correlation matrix does in general not yield a valid correlation matrix. Consider for example three fully correlated scalar values with positive semi-definite joint correlation matrix consisting of only ones. When one of the entries (and its symmetric counterpart) in the correlation matrix is set to zero, the matrix is no longer positive semi-definite.

Lists of Figures, Tables, Algorithms, and Examples

List of Figures

List of Tables

List of Algorithms

List of Examples

Bibliography

[1] S. I. Aanonsen, G. Nævdal, D. S. Oliver, A. C. Reynolds, B. Vallès, et al. The Ensemble Kalman Filter in Reservoir Engineering – a Review. *SPE Journal*, 14(03):393–412, 2009.

[2] J. Ajgl, M. Šimandl, and J. Duník. Millman's Formula in Data Fusion. In *Proceedings of the 10th International PhD Workshop on Systems and Control*, pages 1–6, 2009.

[3] B. D. Anderson and J. B. Moore. *Optimal Filtering.* Courier Dover Publications, 2012.

[4] P. O. Arambel. Structured Pedigree Information for Distributed Fusion Systems. In *SPIE Signal Processing, Sensor Fusion, and Target Recognition XVI*, volume 6968, page 28. Society of Photo-Optical Instrumentation Engineers, 2008.

[5] P. O. Arambel. Performance Analysis of Structured Pedigree Distributed Fusion Systems. In *Society of Photo-Optical Instrumentation Engineers (SPIE) Conference Series*, volume 7336, page 23, 2009.

[6] I. Arasaratnam and S. Haykin. Cubature Kalman Filters. *IEEE Transactions on Automatic Control*, 54(6):1254–1269, 2009.

[7] Y. Bar-Shalom. On the Track-to-Track Correlation Problem. *IEEE Transactions on Automatic Control*, 26(2):571–572, 1981.

[8] Y. Bar-Shalom. Update with Out-of-sequence Measurements in Tracking: Exact Solution. *IEEE Transactions on Aerospace and Electronic Systems*, 38(3):769–777, 2002.

[9] Y. Bar-Shalom and L. Campo. The Effect of the Common Process Noise on the Two-Sensor Fused-Track Covariance. *IEEE Transactions on Aerospace and Electronic Systems*, 22(6):803–805, 1986.

[10] Y. Bar-Shalom, H. Chen, and M. Mallick. One-step Solution for the Multistep Out-of-sequence-measurement Problem in Tracking. *IEEE Transactions on Aerospace and Electronic Systems*, 40(1):27–37, 2004.

[11] Y. Bar-Shalom, P. K. Willett, and X. Tian. *Tracking and Data Fusion.* Yaakov Bar-Shalom, 2011.

[12] G. Battistelli and L. Chisci. Kullback–Leibler Average, Consensus on Probability Densities, and Distributed State Estimation with Guaranteed Stability. *Automatica*, 50(3):707–718, 2014.

[13] M. J. Bayarri and J. O. Berger. The Interplay of Bayesian and Frequentist Analysis. *Statistical Science*, 19(1):58–80, 2004.

[14] R. Bellman. Adaptive Control Processes.

[15] A. R. Benaskeur. Consistent Fusion of Correlated Data Sources. In *Proceedings of the 28th Annual Conference of the IEEE Industrial Electronics Society*, volume 4, pages 2652–2656. IEEE, 2002.

[16] F. Beutler, M. F. Huber, and U. D. Hanebeck. Gaussian Filtering using State Decomposition Methods. In *Proceedings of the 12th International Conference on Information Fusion (Fusion 2009)*, Seattle, Washington, USA, July 2009.

[17] P. J. Bickel and E. Levina. Regularized Estimation of Large Covariance Matrices. *The Annals of Statistics*, 36(1):199–227, 2008.

[18] F. Black and M. Scholes. The Pricing of Options and Corporate Liabilities. *The Journal of Political Economy*, 81(3):637–54, 1973.

[19] V. Blondel and J. N. Tsitsiklis. NP-Hardness of Some Linear Control Design Problems. In *Proceedings of the 34th IEEE Conference on Decision and Control (CDC 1995)*, volume 3, pages 2910–2915. IEEE, 1995.

[20] S. Boyd and L. Vandenberghe. *Convex Optimization.* Cambridge University Press, 2009.

[21] J. Broch, D. A. Maltz, D. B. Johnson, Y.-C. Hu, and J. Jetcheva. A Performance Comparison of Multi-hop Wireless ad hoc Network Routing Protocols. In *Proceedings of the 4th annual ACM/IEEE International Conference on Mobile Computing and Networking*, pages 85–97. ACM, 1998.

[22] J. Capitan, L. Merino, F. Caballero, and A. Ollero. Delayed-state Information Filter for Cooperative Decentralized Tracking. In *Proceedings of the 2009 IEEE International Conference on Robotics and Automation (ICRA 2009)*, pages 3865–3870. IEEE, 2009.

[23] R. Carli, A. Chiuso, L. Schenato, and S. Zampieri. Distributed Kalman Filtering based on Consensus Strategies. *IEEE Journal on Selected Areas in Communications*, 26(4):622–633, 2008.

[24] N. A. Carlson. Federated Filter for Fault-tolerant Integrated Navigation Systems. In *Proceedings of the Position Location and Navigation Symposium (PLANS'88)*, pages 110–119. IEEE, 1988.

[25] N. A. Carlson. Federated Square Root Filter for Decentralized Parallel Processors. *IEEE Transactions on Aerospace and Electronic Systems*, 26(3):517–525, 1990.

[26] F. S. Cattivelli, C. G. Lopes, and A. H. Sayed. Diffusion Strategies for Distributed Kalman Filtering: Formulation and Performance Analysis. In *Proceedings of the Cognitive Information Processing*, pages 36–41. Citeseer, 2008.

[27] F. S. Cattivelli, S. Member, and A. H. Sayed. Diffusion Strategies for Distributed Kalman Filtering and Smoothing. *IEEE Transactions on Automatic Control*, 55(9):2069–2084, 2010.

[28] K. C. Chang, R. K. Saha, and Y. Bar-Shalom. On Optimal Track-to-Track Fusion. *IEEE Transactions on Aerospace and Electronic Systems*, 33:1271–1276, 1997.

[29] A. Charlish, F. Govaers, and W. Koch. Distributed Radar Tracking using the Double Debiased Distributed Kalman Filter. In *2014 IEEE Radar Conference*, pages 1124–1129. IEEE, 2014.

[30] J. Chen and A. H. Sayed. Diffusion Adaptation Strategies for Distributed Optimization and Learning over Networks. *IEEE Transactions on Signal Processing*, 60(8):4289–4305, 2012.

[31] L. Chen, P. O. Arambel, and R. K. Mehra. Fusion under Unknown Correlation – Covariance Intersection as a Special Case. In *Proceedings of the 5th International Conference on Information Fusion (FUSION 2002)*, Annapolis, Maryland, USA, 2002.

[32] C.-Y. Chong and M. Athans. On the Stochastic Control of Linear Systems with Different Information Sets. *IEEE Transactions on Automatic Control*, 16(5):423–430, 1971.

[33] C.-Y. Chong and S. P. Kumar. Sensor Networks: Evolution, Opportunities, and Challenges. *Proceedings of the IEEE*, 91(8):1247–1256, 2003.

[34] C.-Y. Chong, S. Mori, W. H. Barker, and K.-C. Chang. Architectures and Algorithms for Track Association and Fusion. *IEEE Aerospace and Electronic Systems Magazine*, 15(1):5–13, 2000.

[35] C.-Y. Chong, S. Mori, F. Govaers, and W. Koch. Comparison of Tracklet Fusion and Distributed Kalman Filter for Track

Fusion. In *Proceedings of the 17th International Conference on Information Fusion (FUSION 2014)*, 2014.

[36] C.-Y. Chong, E. Tse, and S. Mori. Distributed Estimation in Networks. In *Proceedings of the American Control Conference 1983*, pages 294–300. IEEE, 1983.

[37] S. Cooper and H. Durrant-Whyte. A Kalman Filter Model for GPS Navigation of Land Vehicles. In *Proceedings of the IEEE/RSJ/GI International Conference on Intelligent Robots and Systems (IROS 1994)*, volume 1, pages 157–163, 1994.

[38] L. T. DeCarlo. On the Meaning and Use of Kurtosis. *Psychological Methods*, 2(3):292, 1997.

[39] A. G. Dimakis, S. Kar, J. M. Moura, M. G. Rabbat, and A. Scaglione. Gossip Algorithms for Distributed Signal Processing. *Proceedings of the IEEE*, 98(11):1847–1864, 2010.

[40] A. Doucet, N. De Freitas, and N. Gordon. An Introduction to Sequential Monte Carlo Methods. In *Sequential Monte Carlo methods in practice*, pages 3–14. Springer, 2001.

[41] O. E. Drummond. Track and Tracklet Fusion Filtering. In *AeroSense 2002*, pages 176–195. International Society for Optics and Photonics, 2002.

[42] Z. Duan and X. R. Li. Optimal Distributed Estimation Fusion with Transformed Data. In *Proceedings of the 11th International Conference on Information Fusion (FUSION 2008)*, Cologne, Germany, 2008.

[43] J. Duník, M. Šimandl, and O. Straka. Unscented Kalman Filter: Aspects and Adaptive Setting of Scaling Parameter. *IEEE Transactions on Automatic Control*, 57(9):2411–2416, 2012.

[44] H. Durrant-Whyte. *Introduction to Decentralised Data Fusion*. The Australian Center for Field Robotics, 2004.

[45] M. Emmelmann, B. Bochow, and C. Kellum. *Vehicular Networking: Automotive Applications and Beyond*, volume 2. John Wiley & Sons, 2010.

[46] G. Evensen. The Ensemble Kalman Filter: Theoretical Formulation and Practical Implementation. *Ocean Dynamics*, 53(4):343–367, 2003.

[47] D. Fraenken. *Dynamische Zustandsschaetzung.* Uni Paderborn, 2006.

[48] D. Franken and A. Hupper. Improved Fast Covariance Intersection for Distributed Data Fusion. In *Proceedings of the 8th International Conference on Information Fusion (FUSION 2005)*, volume 1. IEEE, 2005.

[49] J. Friedman, T. Hastie, and R. Tibshirani. Sparse Inverse Covariance Estimation with the Graphical Lasso. *Biostatistics*, 9(3):432–441, 2008.

[50] R. K. Ganti, F. Ye, and H. Lei. Mobile Crowdsensing: Current State and Future Challenges. *Communications Magazine, IEEE*, 49(11):32–39, 2011.

[51] A. Gelb. *Applied Optimal Estimation.* MIT Press, 1974.

[52] G. T. Gilbert. Positive Definite Matrices and Sylvester's Criterion. *The American Mathematical Monthly*, 98(1):44–46, 1991.

[53] I. Gilitschenski and U. D. Hanebeck. Efficient Deterministic Dirac Mixture Approximation. In *Proceedings of the 2013 American Control Conference (ACC 2013)*, Washington D. C., USA, June 2013.

[54] I. Gilitschenski, J. Steinbring, U. D. Hanebeck, and M. Simandl. Deterministic Dirac Mixture Approximation of Gaussian Mixtures. In *Proceedings of the 17th International Conference on Information Fusion (Fusion 2014)*, Salamanca, Spain, July 2014.

[55] F. Govaers, A. Charlish, and W. Koch. Covariance Debiasing for the Distributed Kalman Filter. In *Proceedings of the 16th International Conference on Information Fusion (FUSION 2013)*, pages 61–67. IEEE, 2013.

[56] F. Govaers and W. Koch. Distributed Kalman Filter Fusion at Arbitrary Instants of Time. In *Proceedings of the 13th International Conference on Information Fusion (FUSION 2010)*, pages 1–8, 2010.

[57] F. Govaers and W. Koch. Out-of-Sequence Processing of Cluttered Sensor Data using Multiple Evolution Models. In *Proceedings of the 13th International Conference on Information Fusion (FUSION 2010)*, pages 1–7, 2010.

[58] F. Govaers and W. Koch. Exact Out-of-sequence Processing using the Information Filter. In *Proceedings of the 14th International Conference on Information Fusion (FUSION 2011)*, pages 1–7. IEEE, 2011.

[59] F. Govaers and W. Koch. On the Globalized Likelihood Function for Exact Track-to-Track Fusion at Arbitrary Instants of Time. In *Proceedings of the 14th International Conference on Information Fusion (FUSION 2011)*, pages 1–5, 2011.

[60] F. Govaers and W. Koch. An Exact Solution to Track-to-Track-Fusion at Arbitrary Communication Rates. *IEEE Transactions on Aerospace and Electronic Systems*, 48(3):2718–2729, 2012.

[61] M. Grant, S. Boyd, and Y. Ye. CVX: Matlab Software for Disciplined Convex Programming, 2008.

[62] S. Grime and H. F. Durrant-Whyte. Data Fusion in Decentralized Sensor Networks. *Control Engineering Practice*, 2(5):849–863, 1994.

[63] V. Gupta, B. Hassibi, and R. M. Murray. Optimal LQG Control Across Packet-dropping Links. *Systems & Control Letters*, 56(6):439–446, 2007.

[64] F. Gustafsson, F. Gunnarsson, N. Bergman, U. Forssell, J. Jansson, R. Karlsson, and P.-J. Nordlund. Particle Filters for Positioning, Navigation, and Tracking. *IEEE Transactions on Signal Processing*, 50(2):425–437, 2002.

[65] M. Haenggi. Mobile Sensor-actuator Networks: Opportunities and Challenges. In *Proceedings of the 7th IEEE International Workshop on Cellular Neural Networks and their Applications*, pages 283–290. IEEE, Jul 2002.

[66] U. D. Hanebeck. Kernel-based Deterministic Blue-noise Sampling of Arbitrary Probability Density Functions. In *Proceedings of the 48th Annual Conference on Information Sciences and Systems (CISS 2014)*, Princeton, New Jersey, USA, Mar. 2014.

[67] U. D. Hanebeck, K. Briechle, and J. Horn. A Tight Bound for the Joint Covariance of Two Random Vectors with Unknown but Constrained Cross–Correlation. In *Proceedings of the 2001 IEEE International Conference on Multisensor Fusion and Integration for Intelligent Systems (MFI 2001)*, pages 85–90, Baden–Baden, Germany, Aug. 2001.

[68] U. D. Hanebeck and J. Horn. An Efficient Method for Simultaneous Map Building and Localization. In *Proceedings of SPIE, Vol. 4385, AeroSense Symposium*, Orlando, Florida, USA, Apr. 2001.

[69] U. D. Hanebeck and V. Klumpp. Localized Cumulative Distributions and a Multivariate Generalization of the Cramér-von Mises Distance. In *Proceedings of the 2008 IEEE International*

Conference on Multisensor Fusion and Integration for Intelligent Systems (MFI 2008), pages 33–39, Seoul, Republic of Korea, Aug. 2008.

[70] U. D. Hanebeck and J. Steinbring. Progressive Gaussian Filtering Based on Dirac Mixture Approximations. In *Proceedings of the 15th International Conference on Information Fusion (Fusion 2012)*, Singapore, July 2012.

[71] R. Harle. A Survey of Indoor Inertial Positioning Systems for Pedestrians. *IEEE Communications Surveys and Tutorials*, 15(3):1281–1293, 2013.

[72] J. K. Hart and K. Martinez. Environmental Sensor Networks: A Revolution in the Earth System Science? *Earth-Science Reviews*, 78(3):177–191, 2006.

[73] H. R. Hashemipour, S. Roy, and A. J. Laub. Decentralized Structures for Parallel Kalman Filtering. *IEEE Transactions on Automatic Control*, 33(1):88–94, 1988.

[74] W. K. Hastings. Monte Carlo Sampling Methods using Markov Chains and Their Applications. *Biometrika*, 57(1):97–109, 1970.

[75] J. P. Hespanha, P. Naghshtabrizi, and Y. Xu. A Survey of Recent Results in Networked Control Systems. *Proceedings of the IEEE*, 95(1):138, 2007.

[76] R. A. Horn and C. R. Johnson. *Matrix Analysis*. Cambridge University Press, 2005.

[77] P. Houtekamer and H. L. Mitchell. Ensemble Kalman Filtering. *Quarterly Journal of the Royal Meteorological Society*, 131(613):3269–3289, 2005.

[78] J. Hu, L. Xie, and C. Zhang. Diffusion Kalman Filtering based on Covariance Intersection. *IEEE Transactions on Signal Processing*, 60(2):891–902, 2012.

[79] M. F. Huber and U. D. Hanebeck. Gaussian Filter based on Deterministic Sampling for High Quality Nonlinear Estimation. In *Proceedings of the 17th IFAC World Congress (IFAC 2008)*, volume 17, Seoul, Republic of Korea, July 2008.

[80] M. Ikeda, D. D. Šiljak, and D. E. White. Decentralized Control with Overlapping Information Sets. *Journal of Optimization Theory and Applications*, 34(2):279–310, 1981.

[81] K. Itô et al. Stochastic Integral. *Proceedings of the Imperial Academy*, 20(8):519–524, 1944.

[82] W. James and C. Stein. Estimation with Quadratic Loss. In *Proceedings of the Fourth Berkeley Symposium on Mathematical Statistics and Probability, Volume 1: Contributions to the Theory of Statistics*, pages 361–379. University of California Press, 1961.

[83] A. H. Jazwinski. *Stochastic Processes and Filtering Theory*. Courier Dover Publications, 2007.

[84] H. Joe. Generating Random Correlation Matrices Based on Partial Correlations. *Journal of Multivariate Analysis*, 97(10):2177–2189, 2006.

[85] S. J. Julier and J. K. Uhlmann. A Non-divergent Estimation Algorithm in the Presence of Unknown Correlations. In *Proceedings of the American Control Conference 1997*, volume 4, pages 2369–2373, 1997.

[86] S. J. Julier and J. K. Uhlmann. New Extension of the Kalman Filter to Nonlinear Systems. In *Proceedings of SPIE*, volume 3068, pages 182–193, 1997.

[87] S. J. Julier and J. K. Uhlmann. *Multisensor Data Fusion*, chapter General Decentralized Data Fusion with Covariance Intersection (CI), pages 319–343. CRC Press, 2001.

[88] S. J. Julier and J. K. Uhlmann. Simultaneous Localisation and Map Building Using Split Covariance Intersection. In *Proceedings of the International Conference on Intelligent Robots and Systems (IROS 2001)*, volume 3, pages 1257–1262, 2001.

[89] S. J. Julier and J. K. Uhlmann. Using Covariance Intersection for SLAM. *Robotics and Autonomous Systems*, 55(1):3–20, 2007.

[90] W. Kahan. Circumscribing an Ellipsoid About the Intersection of Two Ellipsoids. *Can. Math. Bull*, 11(3):437–441, 1968.

[91] W. Kahan. Circumscribing an Ellipsoid About the Intersection of Two Ellipsoids, 2006.

[92] J. M. Kahn, R. H. Katz, and K. S. Pister. Next Century Challenges: Mobile Networking for Smart Dust. In *Proceedings of the 5th annual ACM/IEEE international Conference on Mobile Computing and Networking*, pages 271–278. ACM, 1999.

[93] R. E. Kalman. A New Approach to Linear Filtering and Prediction Problems. *Journal of Basic Engineering*, 82(1):35–45, 1960.

[94] R. E. Kalman and R. S. Bucy. New Results in Linear Filtering and Prediction Theory. *Journal of Fluids Engineering*, 83(1):95–108, 1961.

[95] N. E. Karoui. Operator Norm Consistent Estimation of Large-dimensional Sparse Covariance Matrices. *The Annals of Statistics*, 36(6):2717–2756, 2008.

[96] G. Kerschen, K. Worden, A. F. Vakakis, and J.-C. Golinval. Past, Present and Future of Nonlinear System Identification in Structural Dynamics. *Mechanical Systems and Signal Processing*, 20(3):505–592, 2006.

[97] U. A. Khan and J. M. Moura. Distributing the Kalman Filter for Large-scale Systems. *IEEE Transactions on Signal Processing*, 56(10):4919–4935, 2008.

[98] V. Kibardin. Decomposition into Functions in the Minimization Problem. *Automation and Remote Control*, 40(9):1311–1323, 1980.

[99] K. H. Kim. Development of Track to Track Fusion Algorithms. In *Proceedings of the American Control Conference 1994*, volume 1, pages 1037–1041. IEEE, 1994.

[100] W. Koch. On Optimal Distributed Kalman Filtering and Retrodiction at Arbitrary Communication Rates for Maneuvering Targets. In *Proceedings of the 2008 IEEE International Conference on Multisensor Fusion and Integration for Intelligent Systems (MFI 2008)*, pages 457–462, 2008.

[101] A. Kurzhanski and I. Vályi. *Ellipsoidal Calculus for Estimation and Control.* Nelson Thornes, 1997.

[102] O. Ledoit and M. Wolf. A Well-conditioned Estimator for Large-dimensional Covariance Matrices. *Journal of Multivariate Analysis*, 88(2):365–411, 2004.

[103] D.-J. Lee. Unscented Information Filtering for Distributed Estimation and Multiple Sensor Fusion. In *AIAA Guidance, Navigation and Control Conference and Exhibit*, 2008.

[104] T. Lefebvre, H. Bruyninckx, and J. De Schutter. Kalman Filters for Non-linear Systems: a Comparison of Performance. *International Journal of Control*, 77(7):639–653, 2004.

[105] K. Leichtenstern, A. De Luca, and E. Rukzio. Analysis of Built-in Mobile Phone Sensors for Supporting Interactions with the Real World. In *PERMID*, pages 31–34, 2005.

[106] R. Li and V. P. Jilkov. Survey of Maneuvering Target Tracking. Part I. Dynamic Models. *IEEE Transactions on Aerospace and Electronic Systems*, 39(4):1333–1364, 2003.

[107] M. E. Liggins, C.-Y. Chong, I. Kadar, and Et. Distributed Fusion Architectures and Algorithms for Target Tracking. *Proceedings of the IEEE*, 85(1):95–107, 1997.

[108] M. E. Liggins, D. L. Hall, and J. Llinas. *Handbook of Multisensor Data Fusion: Theory and Practice.* CRC Press, 2008.

[109] X. Liu and A. Goldsmith. Kalman Filtering with Partial Observation Losses. In *Proceedings of the 43rd IEEE Conference on Decision and Control (CDC 2004)*, volume 4, pages 4180–4186. IEEE, 2004.

[110] L. Ljung. *System Identification.* Springer, 1998.

[111] J. Luo, D. Joshi, J. Yu, and A. Gallagher. Geotagging in Multimedia and Computer Vision–A Survey. *Multimedia Tools and Applications*, 51(1):187–211, 2011.

[112] A. Makarenko and H. Durrant-Whyte. Decentralized Data Fusion and Control in Active Sensor Networks. In *Proceedings of the 7th International Conference on Information Fusion (FUSION 2004)*, 2004.

[113] A. Makarenko and H. Durrant-Whyte. Decentralized Bayesian Algorithms for Active Sensor Networks. *Information Fusion*, 7(4):418–433, 2006.

[114] G. Marsaglia and I. Olkin. Generating Correlation Matrices. *SIAM Journal on Scientific and Statistical Computing*, 5(2):470–475, 1984.

[115] R. C. Merton. Option Pricing when Underlying Stock Returns are Discontinuous. *Journal of Financial Economics*, 3(1):125–144, 1976.

[116] A. Mutambara. *Decentralized Estimation and Control for Multisensor Systems.* CRC press, 1998.

[117] A. Nedic and D. P. Bertsekas. Incremental Subgradient Methods for Nondifferentiable Optimization. *SIAM Journal on Optimization*, 12(1):109–138, 2001.

[118] W. Niehsen. Information Fusion Based on Fast Covariance Intersection Filtering. In *Proceedings of the 5th International Conference on Information Fusion (FUSION 2002)*, volume 2, pages 901–904. IEEE, 2002.

[119] B. Noack. *State Estimation for Distributed Systems with Stochastic and Set-membership Uncertainties*, volume 14. KIT Scientific Publishing, 2014.

[120] B. Noack, M. Baum, and U. D. Hanebeck. Automatic Exploitation of Independencies for Covariance Bounding in Fully Decentralized Estimation. In *Proceedings of the 18th IFAC World Congress (IFAC 2011)*, Milan, Italy, Aug. 2011.

[121] B. Noack, V. Klumpp, and U. D. Hanebeck. State Estimation with Sets of Densities considering Stochastic and Systematic Errors. In *Proceedings of the 12th International Conference on Information Fusion (Fusion 2009)*, Seattle, Washington, USA, July 2009.

[122] B. Noack, J. Sijs, and U. D. Hanebeck. Fusion Strategies for Unequal State Vectors in Distributed Kalman Filtering. In *Proceedings of the 19th IFAC World Congress (IFAC 2014)*, Cape Town, South Africa, Aug. 2014.

[123] M. NøRgaard, N. K. Poulsen, and O. Ravn. New Developments in State Estimation for Nonlinear Systems. *Automatica*, 36(11):1627–1638, 2000.

[124] R. Olfati-Saber. Distributed Kalman Filter with Embedded Consensus Filters. In *Proceedings of the 44th IEEE Conference*

on Decision and Control (CDC-ECC 2005), pages 8179–8184. IEEE, 2005.

[125] R. Olfati-Saber. Kalman-Consensus Filter: Optimality, Stability, and Performance. In *Proceedings of the 48th IEEE Conference on Decision and Control (CDC 2009)*, pages 7036–7042, 2009.

[126] R. Olfati-Saber, J. A. Fax, and R. M. Murray. Consensus and Cooperation in Networked Multi-Agent Systems. *Proceedings of the IEEE*, 95(1):215–233, 2007.

[127] R. Olfati-Saber and P. Jalalkamali. Coupled Distributed Estimation and Control for Mobile Sensor Networks. *IEEE Transactions on Automatic Control*, 57(10):2609–2614, 2012.

[128] A. Oppenheim. RES.6-007 Signals and Systems, 2011.

[129] T. Pappas, A. J. Laub, and N. R. Sandell. On the Numerical Solution of the Discrete-time Algebraic Riccati Equation. *IEEE Transactions on Automatic Control*, 25(4):631–641, 1980.

[130] K. B. Petersen and M. S. Pedersen. *The Matrix Cookbook*. Technical University of Denmark, 2012.

[131] H. V. Poor. *An Introduction to Signal Detection and Estimation*. Springer, 1994.

[132] J. Postel. User Datagram Protocol, 1980.

[133] J. Postel. Transmission Control Protocol, 1981.

[134] M. Pourahmadi. Joint Mean-covariance Models with Applications to Longitudinal Data: Unconstrained Parameterisation. *Biometrika*, 86(3):677–690, 1999.

[135] M. Pourahmadi. Covariance Estimation: The GLM and Regularization Perspectives. *Statistical Science*, 26(3):369–387, 2011.

[136] M. G. Rabbat and R. D. Nowak. Quantized Incremental Algorithms for Distributed Optimization. *IEEE Journal on Selected Areas in Communications*, 23(4):798–808, 2005.

[137] B. Rao and H. Durrant-Whyte. Fully decentralised algorithm for multisensor Kalman filtering. *IEE Proceedings D, Control Theory and Applications*, 138(5):413–420, 1991.

[138] B. Rao, H. Durrant-Whyte, and J. Sheen. A Fully Decentralized Multi-sensor System for Tracking and Surveillance. *The International Journal of Robotics Research*, 12(1):20–44, 1993.

[139] S. Reece and S. Roberts. Robust, Low-bandwidth, Multi-vehicle Mapping. In *Proceedings of the 8th International Conference on Information Fusion (FUSION 2005)*, 2005.

[140] A. J. Rothman, E. Levina, and J. Zhu. A new Approach to Cholesky-based Covariance Regularization in High Dimensions. *Biometrika*, 97(3):539–550, 2010.

[141] M. Sahebsara, T. Chen, and S. L. Shah. Optimal Filtering with Random Sensor Delay, Multiple Packet Dropout and Uncertain Observations. *International Journal of Control*, 80(2):292–301, 2007.

[142] T. S. Schei. A Finite-difference Method for Linearization in Nonlinear Estimation Algorithms. *Automatica*, 33(11):2053–2058, 1997.

[143] L. Schenato. Optimal Estimation in Networked Control Systems Subject to Random Delay and Packet Drop. *IEEE Transactions on Automatic Control*, 53(5):1311–1317, 2008.

[144] L. Schenato, B. Sinopoli, M. Franceschetti, K. Poolla, and S. S. Sastry. Foundations of Control and Estimation over Lossy Networks. *Proceedings of the IEEE*, 95(1):163–187, 2007.

[145] I. D. Schizas, A. Ribeiro, and G. B. Giannakis. Consensus in Ad Hoc WSNs with Noisy Links–Part I: Distributed Estimation of Deterministic Signals. *IEEE Transactions on Signal Processing*, 56(1):350–364, 2008.

[146] F. C. Schweppe. *Uncertain Dynamic Systems*, volume 160. Prentice-Hall Englewood Cliffs, 1973.

[147] X. Sheng, Y.-H. Hu, and P. Ramanathan. Distributed Particle Filter with GMM Approximation for Multiple Targets Localization and Tracking in Wireless Sensor Network. In *Proceedings of the 4th International Symposium on Information Processing in Sensor Networks*, page 24, 2005.

[148] V. Shin, Y. Lee, and T.-S. Choi. Generalized Millman's Formula and Its Application for Estimation Problems. *Signal Processing*, 86(2):257–266, 2006.

[149] J. Sijs and M. Lazar. Empirical Case-studies of State Fusion Via Ellipsoidal Intersection. In *Proceedings of the 14th International Conference on Information Fusion (FUSION 2011)*, pages 1–8, 2011.

[150] J. Sijs and M. Lazar. State Fusion with Unknown Correlation: Ellipsoidal Intersection. *Automatica*, 48(8):1874–1878, 2012.

[151] J. Sijs, M. Lazar, and P. P. J. v. d. Bosch. State Fusion with Unknown Correlation: Ellipsoidal Intersection. In *Proceedings of the American Control Conference 2010*, pages 3992–3997, 2010.

[152] M. Šimandl and O. Straka. Sampling Densities of Particle Filter: A Survey and Comparison. In *American Control Conference, 2007. ACC'07*, pages 4437–4442. IEEE, 2007.

[153] D. Simon. *Optimal State Estimation: Kalman, H Infinity, and Nonlinear Approaches*. John Wiley & Sons, 2006.

[154] B. Sinopoli, L. Schenato, M. Franceschetti, K. Poolla, M. I. Jordan, and S. S. Sastry. Kalman Filtering with Intermittent Observations. *IEEE Transactions on Automatic Control*, 49(9):1453–1464, 2004.

[155] J. Sjöberg, Q. Zhang, L. Ljung, A. Benveniste, B. Delyon, P.-Y. Glorennec, H. Hjalmarsson, and A. Juditsky. Nonlinear Black-box Modeling in System Identification: A Unified Overview. *Automatica*, 31(12):1691–1724, 1995.

[156] S. T. Smith. Covariance, Subspace, and Intrinsic Crame´r-Rao Bounds. *IEEE Transactions on Signal Processing*, 53(5):1610–1630, 2005.

[157] J. Steinbring and U. D. Hanebeck. Progressive Gaussian Filtering Using Explicit Likelihoods. In *Proceedings of the 17th International Conference on Information Fusion (Fusion 2014)*, Salamanca, Spain, July 2014.

[158] O. Straka, J. Duník, and M. Šimandl. Randomized Unscented Kalman Filter in Target Tracking. In *Information Fusion (FUSION), 2012 15th International Conference on*, pages 503–510. IEEE, 2012.

[159] R. Stratonovich. A New Representation for Stochastic Integrals and Equations. *SIAM Journal on Control*, 4(2):362–371, 1966.

[160] R. L. Streit and T. E. Luginbuhl. Probabilistic Multi-hypothesis Tracking. Technical report, DTIC Document, 1995.

[161] S.-l. Sun. Multi-sensor Optimal Information Fusion Kalman Filters with Applications. *Aerospace Science and Technology*, 8(1):57–62, 2004.

[162] J. Taylor. *Introduction to Error Analysis, the Study of Uncertainties in Physical Measurements.* University Science Books, 1997.

[163] J. K. Uhlmann. Covariance Consistency Methods for Fault-tolerant Distributed Data Fusion. *Information Fusion*, 4(3):201–215, 2003.

[164] D. Van Krevelen and R. Poelman. A Survey of Augmented Reality Technologies, Applications and Limitations. *International Journal of Virtual Reality*, 9(2):1–20, 2010.

[165] R. Vershynin. How Close is the Sample Covariance Matrix to the Actual Covariance Matrix? *Journal of Theoretical Probability*, 25(3):655–686, 2012.

[166] E. A. Wan and R. Van Der Merwe. The Unscented Kalman Filter for Nonlinear Estimation. In *The IEEE Adaptive Systems for Signal Processing, Communications, and Control Symposium 2000*, pages 153–158. IEEE, 2000.

[167] Y. Wang and X. R. Li. Distributed Estimation Fusion Under Unknown Cross-correlation: An Analytic Center Approach. In *Proceedings of the 13th International Conference on Information Fusion (FUSION 2010)*, pages 1–8, 2010.

[168] G. Welch and G. Bishop. An Introduction to the Kalman Filter, 1995.

[169] N. Wiener. Generalized Harmonic Analysis. *Acta Mathematica*, 55(1):117–258, 1930.

[170] D. Willner, C. Chang, and K. Dunn. Kalman Filter Algorithms for a Multi-sensor System. In *Proceedings of the 15th IEEE Conference on Decision and Control (CDC 1976)*, volume 15, pages 570–574. IEEE, 1976.

[171] M. A. Woodbury. Inverting Modified Matrices. *Memorandum Report*, 42:106, 1950.

[172] Y. Xu and J. P. Hespanha. Estimation under Uncontrolled and Controlled Communications in Networked Control Systems.

In *Proceedings of the 44th IEEE Conference on Decision and Control (CDC-ECC 2005)*, pages 842–847. IEEE, 2005.

[173] J. Zhou, Y. Zhu, Z. You, and E. Song. An Efficient Algorithm for Optimal Linear Estimation Fusion in Distributed Multisensor Systems. *IEEE Transactions on Systems, Man and Cybernetics, Part A: Systems and Humans*, 36(5):1000–1009, 2006.

Own Publications

[174] J. Ajgl, M. Šimandl, M. Reinhardt, B. Noack, and U. D. Hanebeck. Covariance Intersection in State Estimation of Dynamical Systems. In *Proceedings of the 17th International Conference on Information Fusion (Fusion 2014)*, Salamanca, Spain, July 2014.

[175] J. Fischer, M. Reinhardt, and U. D. Hanebeck. Optimal Sequence-Based Control and Estimation of Networked Linear Systems. *at - Automatisierungstechnik, Oldenbourg Verlag*, 61(7):467–476, July 2013.

[176] B. Noack, S. J. Julier, M. Reinhardt, and U. D. Hanebeck. Nonlinear Federated Filtering. In *Proceedings of the 16th International Conference on Information Fusion (Fusion 2013)*, Istanbul, Turkey, July 2013.

[177] B. Noack, M. Reinhardt, and U. D. Hanebeck. On Nonlinear Track-to-track Fusion with Gaussian Mixtures. In *Proceedings of the 17th International Conference on Information Fusion (Fusion 2014)*, Salamanca, Spain, July 2014.

[178] M. Reinhardt, S. Kulkarni, and U. D. Hanebeck. Generalized Covariance Intersection based on Noise Decomposition. In *Proceedings of the 2014 IEEE International Conference on Multisensor Fusion and Information Integration (MFI 2014)*, Beijing, China, Sept. 2014.

[179] M. Reinhardt, B. Noack, P. O. Arambel, and U. D. Hanebeck. Minimum Covariance Bounds for the Fusion under Unknown

Correlations. *IEEE Signal Processing Letters*, 22(9):1210 – 1214, Sept. 2015.

[180] M. Reinhardt, B. Noack, M. Baum, and U. D. Hanebeck. Analysis of Set-theoretic and Stochastic Models for Fusion under Unknown Correlations. In *Proceedings of the 14th International Conference on Information Fusion (Fusion 2011)*, Chicago, Illinois, USA, July 2011.

[181] M. Reinhardt, B. Noack, and U. D. Hanebeck. Closed-form Optimization of Covariance Intersection for Low-dimensional Matrices. In *Proceedings of the 15th International Conference on Information Fusion (Fusion 2012)*, Singapore, July 2012.

[182] M. Reinhardt, B. Noack, and U. D. Hanebeck. Decentralized Control Based on Globally Optimal Estimation. In *Proceedings of the 51st IEEE Conference on Decision and Control (CDC 2012)*, Maui, Hawaii, USA, Dec. 2012.

[183] M. Reinhardt, B. Noack, and U. D. Hanebeck. On Optimal Distributed Kalman Filtering in Non-ideal Situations. In *Proceedings of the 15th International Conference on Information Fusion (Fusion 2012)*, Singapore, July 2012.

[184] M. Reinhardt, B. Noack, and U. D. Hanebeck. The Hypothesizing Distributed Kalman Filter. In *Proceedings of the 2012 IEEE International Conference on Multisensor Fusion and Integration for Intelligent Systems (MFI 2012)*, Hamburg, Germany, Sept. 2012.

[185] M. Reinhardt, B. Noack, and U. D. Hanebeck. Advances in Hypothesizing Distributed Kalman Filtering. In *Proceedings of the 16th International Conference on Information Fusion (Fusion 2013)*, Istanbul, Turkey, July 2013.

[186] M. Reinhardt, B. Noack, and U. D. Hanebeck. Reconstruction of Joint Covariance Matrices in Networked Linear Systems. In

Proceedings of the 48th Annual Conference on Information Sciences and Systems (CISS 2014), Princeton, New Jersey, USA, Mar. 2014.

[187] M. Reinhardt, B. Noack, S. Kulkarni, and U. D. Hanebeck. Distributed Kalman Filtering in the Presence of Packet Delays and Losses. In *Proceedings of the 17th International Conference on Information Fusion (Fusion 2014)*, Salamanca, Spain, July 2014.

Karlsruhe Series on Intelligent Sensor-Actuator-Systems

Edited by Prof. Dr.-Ing. Uwe D. Hanebeck // ISSN 1867-3813

Band 1 **Oliver Schrempf**
Stochastische Behandlung von Unsicherheiten in kaskadierten dynamischen Systemen. 2008
ISBN 978-3-86644-287-0

Band 2 **Florian Weißel**
Stochastische modell-prädiktive Regelung nichtlinearer Systeme. 2009
ISBN 978-3-86644-348-8

Band 3 **Patrick Rößler**
Telepräsente Bewegung und haptische Interaktion in ausgedehnten entfernten Umgebungen. 2009
ISBN 978-3-86644-346-4

Band 4 **Kathrin Roberts**
Modellbasierte Herzbewegungsschätzung für robotergestützte Interventionen. 2009
ISBN 978-3-86644-353-2

Band 5 **Felix Sawo**
Nonlinear state and parameter estimation of spatially distributed systems. 2009
ISBN 978-3-86644-370-9

Band 6 **Gregor F. Schwarzenberg**
Untersuchung der Abbildungseigenschaften eines 3D-Ultraschall-Computertomographen zur Berechnung der 3D-Abbildungsfunktion und Herleitung einer optimierten Sensorgeometrie. 2009
ISBN 978-3-86644-393-8

Karlsruhe Series on Intelligent Sensor-Actuator-Systems

Edited by Prof. Dr.-Ing. Uwe D. Hanebeck // ISSN 1867-3813

Band 7	**Marco Huber** Probabilistic Framework for Sensor Management. 2009 ISBN 978-3-86644-405-8
Band 8	**Frederik Beutler** Probabilistische modellbasierte Signalverarbeitung zur instantanen Lageschätzung. 2010 ISBN 978-3-86644-442-3
Band 9	**Marc Peter Deisenroth** Efficient Reinforcement Learning using Gaussian Processes. 2010 ISBN 978-3-86644-569-7
Band 10	**Evgeniya Ballmann** Physics-Based Probabilistic Motion Compensation of Elastically Deformable Objects. 2012 ISBN 978-3-86644-862-9
Band 11	**Peter Krauthausen** Learning Dynamic Systems for Intention Recognition in Human-Robot-Cooperation. 2013 ISBN 978-3-86644-952-7
Band 12	**Antonia Pérez Arias** Haptic Guidance for Extended Range Telepresence. 2013 ISBN 978-3-7315-0035-3
Band 13	**Marcus Baum** Simultaneous Tracking and Shape Estimation of Extended Objects. 2013 ISBN 978-3-7315-0078-0

Die Bände sind unter www.ksp.kit.edu als PDF frei verfügbar oder als Druckausgabe bestellbar.

Karlsruhe Series on Intelligent Sensor-Actuator-Systems

Edited by Prof. Dr.-Ing. Uwe D. Hanebeck // ISSN 1867-3813

Band 14 **Benjamin Noack**
State Estimation for Distributed Systems with Stochastic and Set-membership Uncertainties. 2014
ISBN 978-3-7315-0124-4

Band 15 **Jörg Fischer**
Optimal Sequence-Based Control of Networked Linear Systems. 2014
ISBN 978-3-7315-0305-7

Band 16 **Marc Reinhardt**
Linear Estimation in Interconnected Sensor Systems with Information Constraints. 2015
ISBN 978-3-7315-0342-2